Becoming a CONSTRUCTION MANAGER

› Becoming a CONSTRUCTION MANAGER

John J. McKeon

John Wiley & Sons, Inc.

This book is printed on acid-free paper. ♾

Published by John Wiley & Sons, Inc., Hoboken, New Jersey.

Published simultaneously in Canada.

Library of Congress Cataloging-in-Publication Data:

McKeon, John J.
Becoming a construction manager / by John J. McKeon.
p. cm.
Includes index.
ISBN 978-0-470-87421-9 (pbk.); 978-1-118-11638-8 (ebk); 978-1-118-12273-0 (ebk); 978-1-118-12274-7 (ebk); 978-1-118-12275-4 (ebk); 978-1-118-13020-9 (ebk)
1. Construction industry—Management. 2. Construction management—Vocational guidance.
I. Title.
HD9715.A2M356 2012
624.068—dc23

2011015203

Printed in the United States of America

10 9 8 7 6 5 4 3 2 1

CONTENTS

FOREWORD

OUR "BUILT ENVIRONMENT" shapes almost everything we do and profoundly affects the quality of our lives.

This environment includes the places we live, work, and play; the means by which we travel; the networks that deliver our food, entertainment, healthcare, education, and necessities such as water and electricity.

Once, creating and maintaining these structures was easy. The planners, designers, builders, and end users were the same people. Materials and labor were local, and processes were straightforward.

Today, however, design and construction are extraordinarily complex processes involving the energy and talents of a diverse array of people. Construction has grown into the second-largest industry in the United States (after healthcare), and provides the setting for a wide range of careers.

One of the most challenging, important, and rewarding of these careers is professional construction management.

Professional construction management came into being for the simplest of reasons: Owners—of all kinds of structures, in every industry segment and cost range—felt they could not rely on obtaining good results when they built things. The design and construction process was beset by errors, delays, cost overruns, disputes, and adversarial relationships. The modern era's increasingly complex projects required many different specialists, who often did not communicate or work well together.

Owners, who could never have the scope and depth of expert knowledge these specialists possessed, felt unrepresented and unprotected in a risky and costly enterprise.

The professional construction manager works, on behalf of the owner, to lead a collaborative multi-party team in delivering a high-quality project safely, on time, and on budget.

As our nation strives in the future to rehabilitate and expand many of our most basic infrastructure assets, such as roads, bridges, railways, and airports, this leadership will be ever more important. Similarly, in every type of structure, from theaters to hospitals, from arenas to high schools, CMs will help conceive and execute the best buildings America has ever had.

Becoming a Construction Manager describes the profession of construction management: What CMs do, what knowledge and skills they need, and how excellent professional construction management contributes to our national well-being. In this book, you will hear the voices of actual

CMs, both industry newcomers and seasoned veterans, describing the paths that brought them into the profession and the satisfactions they have found in the field.

You will see, in the book's portfolio of photos and other illustrations, the true breadth of the work CMs help deliver.

This book will introduce you to the origins and growth of the profession, beginning with the recognition that managing a construction project required a distinct set of skills that must be learned and sharpened through both education and experience. You will explore these skills, from the technical aspects of design and construction through the "soft" or "people" skills such as communication, conflict resolution, and leadership.

You will discover the many educational paths that help make a good CM, from undergraduate study to continuing professional training and certification. And you will see how CMs are helping shape a better future through their commitment to goals such as sustainable design and construction and innovation in project financing and long-term operations.

In all of these efforts, the CM's goals are to anticipate, identify, and meet owners' needs for well-planned and well-executed construction projects and programs.

Meeting those goals is critical to our prosperity and to our quality of life. That means that new CMs—including you—will have a central role in creating the communities and opportunities of tomorrow.

BRUCE D'AGOSTINO, CAE, FCMAA
President & Chief Executive Officer
Construction Management Association of America

Becoming a CONSTRUCTION MANAGER

1 What Does a Construction Manager Do?

THE NATIONAL CENTER FOR SUPERCOMPUTING APPLICATIONS at the University of Illinois needed a very special building: Designed and constructed to house the world's fastest supercomputer, the building would require a large machine room with a raised floor, a sophisticated water-cooling system, and an extraordinary electrical load of 24 megawatts. Plus, it needed to withstand tornado winds of 165 miles per hour.

Township Auditorium in Columbia, South Carolina, was on the National Register of Historic Places, and in its 80 years it had hosted performers as diverse as Tony Bennett and the Ringling Bros. and Barnum & Bailey Circus. But the old auditorium had seen better days. A complete overhaul transformed it into a thoroughly modern facility with both state-of-the-art staging resources and excellent amenities, while preserving its historic character.

Nearly nine miles of 12-foot-diameter buried pipe and tunnels, passing through some of America's most sensitive forest areas—that's what it took to deliver a new water distribution system for the Metropolitan Water District of Southern California. Yet the $400 million project was completed under budget and *a full year ahead of schedule.*

National Petascale Computing Facility at the National Center for Supercomputing Applications, University of Illinois. CM by Clayco, Inc. PHOTOS: ©MSTUDIO WEST|MATTHEW MCFARLAND|WWW.MSTUDIOWEST.COM|760.846.2580.

Tunneling to deliver a new water system in Southern California. CM by Hatch Mott MacDonald. PHOTOS COPYRIGHT METROPOLITAN WATER DISTRICT OF SOUTHERN CALIFORNIA.

◀ Sorenson Language and Communication Center, Gallaudet University, Washington, DC. CM by Heery International. PHOTO: ALAIN JARAMILLO.

▼ Exterior view, Sorenson Language and Communication Center. PHOTO: ALAIN JARAMILLO.

Gallaudet University in Washington, DC, serves a unique clientele: deaf and hearing-impaired students. The university's new Sorenson Language and Communication Center is a groundbreaking, one-of-a-kind "visu-centric" learning facility designed so that deaf students can be "seen to be heard."

The project team included deaf participants, and deaf interpreters took part in every meeting, helping to create one of America's most "deaf-friendly" facilities.

The San Francisco Public Utilities Commission is spending some $4.6 *billion* on an expansion and seismic upgrade of the pipelines, dams, reservoirs, and water treatments plants that serve 2.5 million Bay Area businesses and residents. The full program will unfold over a period of nearly 20 years.

Ford's Theater in Washington, DC, is one of America's most famous and historically significant buildings, and the National Park Service wanted to complete a comprehensive rehab to mark Abraham Lincoln's 200th birthday in 2009. Extensive changes were needed to bring Ford's into compliance with the Americans with Disabilities Act. In addition, there were extensive modernizations to the lighting, acoustics, theater seating, audiovisual systems, HVAC, and other systems . . . all conducted while maintaining the building's historical integrity. The Woodrow Wilson

Historic Ford's Theater as renovated for Abraham Lincoln's 200th Birthday observance in 2009. CM by CH2M HILL. PHOTOGRAPHY BY KENNETH M. WYNER.

Bridge near Washington, DC, was one of the most complex and ambitious construction programs in the country, replacing an antiquated drawbridge with a pair of new, 12-lane bridges. Yet, through its construction timetable of more than a decade, it remained on budget and on schedule. This program involved a remarkable array of innovations both in construction methods and in management of contracts and interfaces. The real key to its success, however, was the *collaboration* that characterized the large, complex, multi-player project team.

All of these projects succeeded in large part because they made use of professional construction managers.

◀ ▼ Replacing the old Woodrow Wilson Bridge near Washington, DC, involved interaction among four governments, 32 prime contracts, and more than 200 subcontractors. The $2.4 billion, 11-year project was led by Parsons Brinckerhoff through the joint venture Potomac Crossing Consultants. PHOTOGRAPHS BY JOSEPH ROMEO PHOTOGRAPHY (LEFT) AND EYECONSTRUCTION (BELOW).

Construction management is a relatively young profession that has already had a profound impact on how America plans and constructs its built environment. That influence is certain to expand as the future brings new imperatives, from tighter budgets to greater environmental sustainability.

Construction management is the practice of professional management applied to the planning, design, and construction of projects from inception to completion for the purpose of controlling time, scope, cost, and quality.

Construction managers (CMs) plan, direct, coordinate, and budget a wide variety of construction projects. According to the U.S. Department of Labor, these projects include residential, commercial, and industrial structures; roads and bridges; wastewater treatment plants; and schools and hospitals. CMs schedule and coordinate all design and construction processes, including the selection, hiring, and oversight of specialty trade contractors such as carpenters, plumbers, and electricians.

A CM rarely does the actual construction work; he or she is the overall manager, the glue that holds the job together. CMs supervise the construction process from conceptual development through final construction, making sure that the project gets completed on time and within budget. They often work with property owners, developers, engineers, architects, and others who are involved in the process.

Depending upon their organization, CMs are sometimes called project managers, construction superintendents, project engineers, or construction supervisors. These titles can be misleading and inaccurate, however, because the scope of work CMs take on—and the education, experience, and skills they possess—go beyond these titles, which also can be used for other positions.

In some instances, the CM may work for the project owner or developer, or for a construction company.

If a project is so large that one person cannot manage it—such as an office building or an industrial complex—duties may be divided among several CMs, mainly based on phases of the job. For example, different CMs may be in charge of (1) site preparation, including clearing and excavation of the land, installing sewage systems, and landscaping and road construction, (2) building construction, including laying foundations and erecting the structural framework, floors, walls, and roofs, and (3) building systems, including protection against fire and installing electrical, plumbing, air-conditioning, and heating systems.

What Makes CM a Profession?

Many CM practitioners would recoil from hearing their work described as "paraprofessional." Yet that's exactly the word used by a noted academic in a 2010 guest column in *Engineering News-Record* (October 25 column, p. 64):

"Trends indicate that the role of CM eventually could be considered simply a paraprofessional who works under the direction of an individual with a state-regulated professional license," the professor wrote.

This column appeared at a time when several initiatives were moving forward simultaneously, all based on determining exactly what "professionalism" means as applied to CM and what education, skills, and credentials the industry should rely on in identifying true CM pros.

A joint meeting of the Board of Directors of the Construction Management Association of America (CMAA) and the Board of Governors of the Construction Manager Certification Institute (CMCI), held in San Diego in 2010, focused on these topics.

Defining a Profession

"The CM profession is largely unregulated," the group noted. Some participants held that "One key to the definition of a profession is that it is licensed by somebody. Every profession is licensed." However, it is common for CM to be performed at a high level by individuals holding licenses as engineers or architects, or holding no government-issued license at all.

If a license is not crucial to the definition of a professional, what elements are?

Andrew Abbott, in his book *The System of Professions: An Essay on the Division of Expert Labor* (University of Chicago Press, 1988), defines a profession as "an occupational group with some special skill." Among the keys to a profession's success is its ability to claim jurisdiction over particular tasks and to determine who is entitled to perform those tasks.

The rise of university education has reinforced this power. Abbott holds that universities have served as "legitimators of professional knowledge and expertise." In the same spirit, the joint Board meeting found that "CMs today are better educated than in the past, and their scope of work is larger than before, including more social issues. The academic course of study is much more recognized today, and there are more programs than ever before."

Indeed, CM-specific university programs are becoming both more numerous and more accepted. One reflection of this trend is CMCI's recent decision to recognize, in reviewing applications for the Certified Construction Manager (CCM) program, relevant undergraduate degrees granted by any institution accredited under the Council for Higher Education Accreditation. In the past, this recognition has been extended only to degrees granted by programs accredited by the Accreditation Council for Construction Education (ACCE) or the Accreditation Board in Engineering Technologies (ABET).

Although educational resources continue to grow, a critical element of true professionalism, according to Abbott, is the fact that the profession itself determines what skills and knowledge it requires of newcomers. To be successful, a profession must be able to control who can claim membership. In this regard, Abbott says, "a single, identifiable national association is clearly a prerequisite."

The Professional Process

Another key to the definition of a profession is the standard to which a practitioner is held in assessing the outcome of his or her work. The CMAA College of Fellows addressed this topic in its 2010 white paper, *Managing Integrated Project Delivery.* The basic question, the Fellows noted, is whether the CM is expected to deliver a *product* or to manage a *process.*

If, by doing a job in a workmanlike way, a practitioner can be expected to deliver the same product to the same specifications every time, then the standard by which the work is judged is the creation of a "defect-free" product.

On the other hand, in some cases every job is different, and conditions may be difficult to predict, let alone control. In such cases, the Fellows' paper says, "professionals . . . are in the business of exercising learned judgment, based on experience with a body of knowledge and on situations and decisions not totally knowable or under their exclusive control."

The U.S. Office of Personnel Management (OPM) makes the same point in different words in its Job Family Series descriptions:

> Professional work involves exercising discretion, analytical skill, judgment, personal accountability, and responsibility for creating, developing, integrating, applying, and sharing an organized body of knowledge:
>
> - Uniquely acquired through extensive education or training at an accredited college or university;
> - Equivalent to the curriculum requirements for a bachelor's or higher degree with major study in, or pertinent to, the specialized field; and
> - Continuously studied to explore, extend, and use additional discoveries, interpretations, and applications to improve data quality, materials, equipment, applications, and methods.

These two descriptions show that definitions of professionalism tend to focus on three elements: (1) discretion or judgment in executing the work, based on (2) specialized education and (3) an organized, broadly accepted body of knowledge.

Performance Standards and Increasing Scope

Because of the prominence of individual judgment in delivering professional services, and the high variability of job conditions, professionals are measured by a metric different from that of "defect-free" products—namely, a "standard of care."

Under this doctrine, each professional is expected to deliver "the same level of care employed by reasonably prudent professionals practicing in the same field in the same area."

KEVILLE ENTERPRISES, 2011, PHOTO BY RICH SARLES.

The professional's ability to adapt to changing and unpredictable circumstances is becoming more critical, the combined Boards of the CMAA and the CMCI noted. "Often the technical part of a project is a slam dunk, but a large part of the job is on the social side." This observation reflects a key expansion of the CM's role in recent years.

Increasingly, CMs today are called upon to manage stakeholder relations, sustainability issues, and a whole gamut of other tasks calling for mastery of so-called "soft skills." The combined Boards noted that business acumen is often as critical to a CM today as technical ability, and effectiveness in financial management, communication, and leadership often makes a decisive contribution to project success.

Similarly, today's CM is often called upon to advise owners about many aspects of a project that fall outside of the traditional construction stages. Technical and "soft" skills are important, but so is the CM's adherence to an established code of ethics that is recognized by all practitioners.

Credentialing Professionalism

Professional credentials—as long as they are recognized as being fairly earned through demonstrated mastery of an accepted body of knowledge—can serve many of the same purposes as a license. Continuous improvement and enlargement of the body of knowledge through research and practical experience are essential to gaining this recognition.

The International Organization for Standardization (ISO) Standard 17024 provides the means by which personnel certification programs can ensure their transparency, consistency, and equity. The American National Standards Institute is authorized by ISO to accredit certification programs that comply with ISO 17024, and has granted this accreditation to the CCM program.

Yardsticks of Professionalism

In summary, a profession can be identified by these qualities:

Its practitioners exercise independent judgment in response to conditions that vary and are not easily predicted or controlled.

This judgment is shaped by rigorous and specialized education, coupled with a broadly accepted system of standards of practice.

The standards of practice and related body of knowledge undergo continuous review and improvement.

The professional practice is governed by a code of ethics and represented by a single nationwide association.

The Big Picture

CHARLES B. "CHUCK" THOMSEN, FCMAA, FAIA

Why did you want to become a construction manager?

› I'm an architect and I just got very interested in seeing things built, getting them done and speeding up the process, and seeing a tangible result in the world, not living in a paper world.

What kind of work do you do?

› I'm retired now, so I consult. But, over my career we managed construction for lots of projects around the world, 20 countries. We renovated the Pentagon; we did quite a bit of work on the Los Angeles school system. We did lots of traditional government work and schools, and our company also did architecture and engineering.

What would you say the most important skills are for a construction manager?

› A clear view of the process, technical skills to deal with the details of design and construction, and the ability to work effectively with people.

For you, what has been your greatest challenge?

› Always, in running a company, you have the continuous challenge of ensuring you are doing good work, keeping really good people effectively engaged working together collaboratively, educating clients, understanding their needs, and maintaining that two-way communication.

What is your educational background?

› I have an undergraduate degree from University of Oklahoma, an intermediate year at the University of Minnesota, and then a graduate degree from MIT. Both degrees are in architecture.

What are some of the working conditions like for a construction manager?

› Well, it depends on what you do. You might sit in an office, or you might sit in a trailer on a construction site; you might be in the United States in an elegant office building in a downtown urban environment, or you might be in the deserts of the Mideast or in a rainforest, or whatever field conditions or office conditions, or both.

Tell me about some of the people who have influenced your work and your profession.

› The senior officers of the companies I worked with, both when I was young and learning, and who provided leadership to me, and then the bright young men and women who worked with me when I got to be chairman and president and a leader within the profession. And of course in that process you always meet some clients who are very influential.

In general, what challenges do you think the profession has ahead of it?

› The construction industry is extraordinarily large, I guess the largest in the world, and the second largest in the United States; healthcare is bigger. Buildings are extraordinarily complicated, and they are designed and fabricated by hundreds of organizations before you get the project done. They are regulated by a handful of organizations; the owners frequently have an administrative group, a user group, and technical managers to manage the facility. Having all that brainpower plugged into the project at one time is the challenge we face now and will continue to face. You have to get those folks working together effectively at the right time. You still have to think about all those hundreds of organizations and the thousands of people involved.

There is an enormous amount of brainpower, and bringing all that brainpower together at the right time and in the right way so they can all make their contribution to the project effectively I think is the biggest challenge.

What advice do you have for students?

› Most schools of construction science in the United States essentially think that they are training students to be general contractors. In the process of doing that, they talk about the technical aspects of building a building, how you make a cost assessment, how you schedule design and construction, how you manage homework, how you manage a safety program—all those kinds of things. They don't spend so much time talking about the complete process of managing design and construction. They're talking about how you get one brick on top of another, and not how you manage those thousands of people.

I would think you might want to talk about the difference between being a construction manager and being a general contractor. There is an awful lot of gray area there, and a lot of people would argue they are the same thing, but there is a difference in interests. There is what you do during the design phase, and then what you do to coordinate everybody during construction. Well, if you look at the history of the country or the industry, you'll see that particularly in the earlier part of the nineteenth century, builders call themselves architects and there wasn't much difference, really. It wasn't until the middle of the century that architecture and engineering separated from construction, moved in out of the mud and the rain and began to design buildings that were beautiful and functional. At that time, architects were the brains, they had the technical knowledge, and contractors were the brawn. They built what they were told to build by the architects and engineers. Schools then started teaching architecture, and we started being selected on the basis of qualifications rather than selling a product. We were selling a service like doctors and lawyers.

In the late twentieth century, the same thing happened to the management of construction. There was so much specialization in some of these subcontractors that the builders who used to lay the brick and put up the timber suddenly were dealing with 75 subcontractors, and the challenge became management rather than putting one brick on top of the other. Just like architects formed AIA, managers of construction formed CMAA. We got schools that teach management design and construction, just like MIT formed the first school of architecture in 1863. Now they are being selected on the basis of qualification, just like architects. History is repeating itself, and schools need to be conscious of this and need to be conscious that the management of the entire process of the design and construction is not just about the technology of building.

Infrastructure projects like this bridge in Greenville, Mississippi, are a prime arena for professional CM. PHOTO CREDIT: JANET WARLICK CAMERA WORK, INC.; COURTESY OF HNTB, INC.

Construction Managers Are Logicians, Hirers, and Organizers

CMs determine the best way to get materials to the building site and the most cost-effective plan and schedule for completing the project. They divide all required construction site activities into logical steps, estimating and budgeting the time required to meet established deadlines. Doing this may require sophisticated scheduling and cost-estimating techniques that rely on computers with specialized software.

CMs also manage the selection of general contractors and trade contractors to complete specific phases of the project—which could include everything from structural metalworking and plumbing to painting, installing electricity, and laying carpeting. CMs determine the labor requirements of the project. They sometimes oversee the performance of all trade contractors and are responsible for ensuring that all work is completed on schedule. CMs direct and monitor the

Professional construction managers bring their organizational and team-building skills to large, complex projects like the New Meadowlands Stadium in New Jersey, home of the NFL's Giants and Jets. PHOTOS COURTESY OF SKANSKA USA.

progress of construction activities, occasionally through construction supervisors or other CMs. Sometimes they are responsible for obtaining permits and licenses and, depending upon the contractual arrangements, for directing or monitoring compliance with building and safety codes, other regulations, and requirements set by the project's insurers. They also oversee the delivery and use of materials, tools, and equipment; worker safety and productivity; and the quality of the construction.

In short, CMs are the overseers of construction projects from start to finish. The buck stops with them.

A Summary of Construction Manager Tasks (from the U.S. Department of Labor)

- Schedule the project in logical steps and budget time required to meet deadlines.
- Confer with supervisory personnel, owners, contractors, and design professionals to discuss and resolve matters such as work procedures, complaints, and construction problems.
- Prepare contracts and negotiate revisions, changes and additions to contractual agreements with architects, consultants, clients, suppliers, and subcontractors.
- Prepare and submit budget estimates and progress and cost tracking reports.
- Interpret and explain plans and contract terms to administrative staff, workers, and clients, representing the owner or developer.
- Plan, organize, and direct activities concerned with the construction and maintenance of structures, facilities, and systems.
- Take actions to deal with the results of delays, bad weather, or emergencies at construction site.
- Inspect and review projects to monitor compliance with building and safety codes, and other regulations.
- Study job specifications to determine appropriate construction methods.
- Select, contract, and oversee workers who complete specific pieces of the project, such as painting or plumbing.

Right to the Point

ROBERT WILSON, FCMAA, RETIRED

What is your background?

› I was with the Navy Civil Engineer Corps, and basically did construction management for the Navy. When I retired from the Navy in 1979, I went to work for Gilbane Building Company and did the same thing. Then I also worked for a while with Parsons Brinckerhoff doing construction management.

What's your education?

› I have a bachelor's in mechanical engineering and a master's in petroleum engineering.

What attracted you to construction management?

› When I was in the Navy Civil Engineer Corps, I just gravitated to CM.

What advice would you give someone who would like to be a CM?

› Learn scheduling. One of the things I was never interested in was English and writing, and that is extremely important. Those are the two main pieces of advice, and I would also add, join the professional societies and participate. It makes a big difference.

In your career in construction management, what was your greatest challenge?

› People. Sometimes they're like herding cats. To keep everyone working toward the same objective and being totally aboveboard and fair is not always easy.

What is the most satisfying part of being a construction manager?

› A completed project where everybody is happy with the result.

And the least satisfying part?

› When you end up in litigation.

Has that happened to you?

› Yes.

Is that just part of the turf? Is it kind of like being a surgeon and just expecting that once or twice a year, you're going to get sued?

› It used to be more that way than it is now. There is more emphasis on partnering, on all of the newer delivery construction techniques, where people work together to achieve the same objective, and negotiate contracts rather than low-bid contracts.

That lowers the amount of litigation?

› Yes. Definitely. Absolutely.

What should a person know about being a construction manager?

› Understanding people and understanding what motivates them is probably as important as a lot of other things. It's really a people business.

Right, so it speaks to the idea that you can't just be a civil or some other engineer, but you really need to understand the soft skills.

› Yes, the personnel skills or people skills, or whatever you want to call it.

Is there any project of yours that sticks in your mind?

› Our most satisfying one was the national World War II Memorial. It was a long overdue project. It overcame a lot of hurdles in the regulatory process. It's constructed on the National Mall. The WWII veterans were dying out relatively fast when we were building it, and we spent a long time in the design phase. Because of that, the motivation of everybody involved was to do what it took to get it done, and that really showed up in the construction phase. Every worker was proud of what they were doing, and that is very unusual. We planned it for two and a half years, and I think it took two months longer than that. It really came off well.

History of Construction Management

For centuries, most construction projects were directed and executed by a single person or team, who commanded all of the skills and knowledge necessary for the job. Thomas Jefferson, the third president of the United States and principal author of the Declaration of Independence, was an example. For his now-famous home at Monticello in Virginia, Jefferson served as architect, construction supervisor, and procurement officer, and even oversaw the making of bricks on his property.

As construction projects grew increasingly complex, however, no individual project owner, architect, contractor, or other participant brought that many skills and areas of knowledge together. Separate organizations advanced the interests of each of these specialties. In directing and balancing the interests of all these participants, including some turf wars, the owner faced a daunting task, which was made more difficult by the desire for speed in moving projects from concept to completion.

These trends had severe impacts on the construction industry. Other factors made construction management almost inevitable. One of the main reasons for the emergence of construction management was the double-digit inflation in the United States in the mid-1960s. Builders were deterred from taking on projects, especially large ones, or were abandoning them before completion. Owners were frustrated, never knowing how much their projects would cost, how long they would take, or if they could be completed at all. The industry needed to find a cheaper, more efficient way to deliver a completed project on time and on budget.

One part of the solution, clearly, was to find a way of executing projects more quickly. The response was so-called "fast track" construction, a system that allowed different portions of the traditional construction work flow to proceed concurrently. For example, portions of a project would be put out for bid before the design for the entire project was completed.

Fast track construction brought with it an acknowledged higher risk of errors in design, execution, or both. The proliferation of specialists involved in even relatively straightforward projects, coupled with the intense economic pressure on owners, created a clear and resounding need for leadership. That leader would be the CM.

According to Chuck Thomsen, one of the authors of a pivotal 1975 joint statement by the American Institute of Architects, the Associated General Contractors, and the American Consulting Engineers Council, identifying CM as a distinct professional discipline, the inflation of 15 to 18 percent annually during the 1960s meant that the buying power of every budgeted dollar declined at a breathtaking rate. Public agencies were particularly affected by this trend, since so much public construction was funded by bond issues, and the gross amount of money available for any given purpose could not easily be increased. Because of these money constraints, the construction industry had to become more efficient, and that meant having someone at the top management level watching all costs and making sure every penny was spent wisely. Again, this all pointed to one person: the CM.

Working for the Government's Real Estate Company

CHARLES HARDY, CCM

Director

U.S. General Services Administration

Tell me how you got into the industry.

› We are pretty much a real estate arm of the government, so we act like a developer, if you will. We also construct and manage new buildings and renovation projects within those buildings. So we are touching all aspects of it. I am an architect by education and training, and I was out in the private sector doing architectural work and then just made that transition over to the owner side, using those same skill sets and broadening them to a much more cradle-to-the-grave situation.

What attracted you to government service?

› I am a retired Air Force officer. I have always seen civil service as something people should do. It's not just the civic responsibility; it's the breadth of the program as well. We are the largest real estate developer in the country and, some could argue, the world. The level of projects we do is second to none and it is an exciting job.

What is the difference between being a CM and doing what you're doing in government service as opposed to the private sector?

› They're variations on a theme. We certainly have our rules and regulations laws and bureaucracy that we deal with, but the same could be said about the outside, when you look at the ways companies are set up and structured. So while they are probably different at face value, in the end, whether it is the public sector or the private sector, there are certain constraints you have to deal with, and you work with those. One of the biggest differences between the public sector and the private sector is there are things that the public sector, as the federal entity, will do for the greater good that may or may not make business sense to a burn-and-churn kind of property owner that will pump a building in 10 years. Our buildings are held for 50, 100 years, or as long as the republic is going on. So there is a much more longer-term perspective than on the public side. The private side is the in-and-out kind of perspective; the public side is a lot more strategic.

On a day-to-day level, how does that affect what you do?

› First costs are not always necessarily the focus. The long-term maintenance issues are brought into it; the quality of materials, the impact it has on the operation maintenance of the facility—all those kinds of things—start entering in much more strongly than they would on a regular project development.

How do you characterize what's going on in the federal government as far as recognition for construction management is concerned?

› I think it's really paralleling and aligned with the industry right now. We're seeing much more highlight on the construction management industry as a whole. Colleges and programs are getting more robust, and people are no longer targeting themselves an architecture or engineering degree. There are construction management degrees.

The government always tends to have such a large influence and to be heard across the country; when we start taking actions, it has to help move the industry in those directions. My sense is this push that we're having here will take construction management to the next level where it needs to go and become more mainstream.

Is there any advice you would give somebody wanting to enter the field of construction management?

› Construction management touches on all aspects of construction, so if you are kind of interested in the industry but not quite sure where you want to go, construction management provides that. And if you do know where you want to go, it still provides you that broad perspective, so when you do go specialize later into architecture or into engineering or into general contracting, you have kind of that holistic view of what you are doing. It's a good foundational arm, or if you are in an alternate career and you start in architecture like I did and move into it, it's another one that helps you have more impact on the industry. It's just an all-around good profession.

Architects, Contractors, and Engineers Weigh in on Construction Management

The profession of construction management was formally acknowledged by the American Institute of Architects, Associated General Contractors, and American Consulting Engineers Council in a joint position statement in 1975. That statement said:

> A background in construction contracting, architecture contracting and engineering can provide a basis of experience for entering the field of Construction Management. However, the basic minimum capabilities of contractors, architects and engineers do not necessarily or automatically provide an individual with all of the skills required of a competent CM.

> From a practical standpoint, an effective CM organization is likely to be a multi-discipline organization. However, CM is an appropriate function for construction contractors, as well as architectural or engineering firms or divisions thereof, so long as said organization or division, in fact, has CM capabilities.

Another construction management pioneer, George Heery, who also helped draft the 1975 joint statement, noted that his firm, Heery & Heery, developed a means of fast-tracking major projects in response to client concerns about the impact of inflation on the cost of long-lead-time items. For example, a major manufacturing facility built in the late 1960s for Lockheed was to support the construction of the C5A cargo jet. "We figured out what the long lead items were, such as the structural frame," Heery says. "We worked it out so that Lockheed bought those things with us managing the process. We didn't award the general contract until we had the final drawings and specifications finished. That was how we did most of our early highly accelerated projects."

Again, the person who could see the big picture, and understand a complex project from start to finish, was the CM.

Indeed, complexity was another key factor in the growth of construction management. According to *Construction Management: A Professional Approach*, by Thomas C. Kavanagh, Frank Muller, FCMAA (more on these titles later), and James J. O'Brien, FCMAA (McGraw-Hill, New York, 1978), construction management gained a foothold in the industry by the mid-1970s. The book, widely regarded as one of the first true textbooks of construction management, noted: "Construction management offers a fresh approach to filling the gaps between construction, engineering and management as the concepts of the institutions, traditions and, in essence, the basic fabric of the construction industry evolve with changing needs. . . . Future projects will be characterized by a shift to large-scale undertakings, large organizations, and a telescoping of the traditional architect/engineer-contractor relationship in a team attack, aided by social, behavioral and environmental scientists, to meet the needs of society."

This was echoed by *Building Construction: Principles, Materials, and Systems* (Pearson-Prentice Hall, New Jersey, 2008), by Madan Mehta, Walter Scarborough, and Diane Armpriest. They noted:

> In the 1970s, because of the large cost overruns and time delays on many projects, owners began to require architects to include a cost estimating professional in the early stages of the design process. . . . [This] often meant involving the contractor during the design phase. As the contracting community acquired the ear of the owner during the design phase, it began to influence many issues that in the traditional method were entirely within the architect's realm. The contractor's involvement in the design phase, the increased complexity of building projects, and the owners' push for timely and on-budget project deliveries made the contractors realize the need for professional management assistance in construction. This gave birth to the full-time professional Construction Manager.

In summary, among the most critical factors leading to the emergence of professional construction management were owners' disenchantment with cost and schedule overruns and the growing recognition that projects would only become more complex in the future, demanding skills that did not currently exist in the traditional construction environment.

Although the terms construction management and CM were beginning to be used in the mid- to late 1960s, *Professional Construction Management*, by William Foxhall, published in 1972 by McGraw-Hill, appears to be the first book devoted to the subject.

A Family Tradition

ROCCO VESPE, PE, FCMAA

Director of the Philadelphia office of Trauner Consulting Services

What got you interested in being a construction manager?

› My family had a construction business. When I first got out of high school, I did some work in construction. It was as a result of my brother and me working in construction, that we started a construction business. For the first, maybe 15 years of my career, I was actually in the construction contracting business; we did concrete structures, buildings, bridges, and all kinds of different projects around the tristate (New York, New Jersey, Connecticut) area. Then I switched over to the consulting end. I applied what I had learned as a contractor and a project manager for a contractor to provide construction management services for other parties. What do I like about it? I like the fact that you're able to build or create something that you're in charge of, like taking land and converting it into some type of structure or some type of facility. Certainly, there are lots of different self-satisfying things that I enjoy in terms of being able to keep people focused on an objective. I still do that in the work that I do now in terms of managing projects and also helping people to deal with problems that come up on problem projects.

What is your formal training?

› I'm a degreed engineer from Drexel, and I'm a licensed professional engineer in seven states. Also, I'm a Fellow of the CMAA. Pretty much my whole career has been in construction.

Were there other major influences on you that made you move in this direction?

› My education was in civil engineering and construction management. My education was really kind of chosen because we were in that business. My brother had a family, and he couldn't go back to college, so I went back to college. That was the way it turned out. One of us needed to learn more about the business, industry, and skills needed in order to be a good contractor and construction manager. There are a lot of similar skills needed.

What are your responsibilities now?

› I'm a director at Trauner Consulting Services. I manage, mentor, and perform lots of different services. Everything ranging from expert testimony in litigation, mediation, and arbitration associated with construction, seminars, training, project management on ongoing projects, project and construction management on troubled projects, claims avoidance and claims review (a claim being a dispute among owners and contractors), and some writing and research. I brought some courses to the National Highway Institute on partnering and claims avoidance, and I go out and teach those courses. Pretty much lots of services related to construction, and the risk, and the trials and tribulations that come about from someone deciding to build something. Also, I've written before. I wrote a book on construction estimating that was published by McGraw Hill [*Construction Estimates from Take-Off to Bid*]. It was a while ago, probably 15 years ago. It was one of the first things Ted Trauner, the principal here, got me involved with. He and I basically updated and revised a text on estimating.

What do you think are the greatest challenges for people getting into construction management?

› From my own experience, it is the ability to apply what you've learned in college, as an undergraduate and a student from books. Apply that to the real world in terms of the factors that occur and really drive a construction project that you may or may not have been able to learn about in school. In other words, how do you apply that to the real world? You can study scheduling, for example, you can learn how to do scheduling. But, in order to be a good scheduler, you have to know how a building goes together. The best way to know how a building goes together is having been involved in building a building. It's a matter of getting the real feel of hands-on experience that goes with the technical expertise and education. I think that is where the challenge is.

In other words, taking the theoretical and moving it over to the real world?

› Yes. Most construction projects are a combination of people out in the field that know how to do the work, that may not have an education or the technical background, other than their own field, and leaders that have been to school and have studied how to do the controls and the scheduling and so on. You need both. A good project manager is somebody who has knowledge on both sides. That can see it from the guy out in the field's perspective and also see it from the technical perspective.

For you personally, what has been your personal greatest challenge in this business?

› For me, probably the biggest challenge was the challenge of transitioning from a project manager/contractor to a consultant. Let me explain it this way: As a project manager, often you have to make decisions because you need to keep the project moving. Sometimes you are making decisions on 75–80 percent of the information you would really like to have in order to make that decision. You look at the decisions you made in the course of the day. If you made twenty, 18 of them were good, and the other two you can sort of fix up the next day. As a consultant though, when somebody is hiring you to give an opinion, write a report, or make a conclusion, you really want to make sure you've looked at everything that's applicable or relevant to the decision you are making. Not until you've looked at all that should you reach a conclusion or make a decision. So, there is a different perspective, and one of my biggest challenges was making that transition from project manager to a professional consultant.

Do you miss the action?

› No, because I do some of both. I have a couple ongoing projects now and a couple ongoing disputes and claims that I'm working on, and a couple reports that I'm writing. So I have a nice mixture of work, and that's one of the reasons why I keep working. I enjoy that.

Was there any particular project or job that you found particularly satisfying?

› Let's start with the ones that I thought were particularly satisfying. I did a major bridge project up in Maine that was a pretty significant challenge. It was a $125 million project, it was four phases, and we were involved from the beginning in terms of preparing specifications, general conditions, and all that information. We were involved in the monitoring, and the project turned out really well.

What was so particularly satisfying about it?

› Just being able to really be a part of a team that not only identified problems up front, but also dealt with the problems as they were going on, and being able to get over the problems without knocking the train off the tracks. When the train gets off the tracks it's sometimes very difficult to get it back on.

Then the opposite question: Is there any project that you are just glad to be done with?

› I can't think of one just right off the top of my head that would be a good example. Sometimes, we are brought into troubled projects, but even there it's a challenge to get it done so it wouldn't be the worst project I got involved with. It would be maybe a good project from the standpoint that I was able to resurrect it and get it back online.

There was one person who told me today that there was one project where the people were just very annoying, would change their minds, and couldn't make decisions. It was very frustrating. He realized it was part of the job, but this case in particular was a little vexing.

One of the projects I just recently finished was a casino project north of Philadelphia. The casino business is known for changes. Fortunately, we were the owner's construction manager, but we had a contractor that was used to doing casinos and used to dealing with changes. They just figured out ways to overcome the changes because that's what they do. It's like anything else. If you hire someone that represents themselves as building casinos all the time, I would say to them, "Don't complain about the changes, this is something you told me you had experience in being able to figure out ways to work around." And they did, they did a good job. We got opened on time, under budget, beautiful building. It was a satisfying experience. That was probably my second-favorite project. That was another that was $180 million.

Where do you think the profession will be going forward from here on out? What changes do you see?

› Probably the most significant change I see in the industry is the advent of building information modeling (BIM). It is going to have an effect on one of the ingredients that is a source of problems on many projects. That is design problems as they relate to interferences, not enough space, and human error that comes along with having to assemble, gather, draw, and organize the amount of data that is necessary to build a building. With building information modeling, you're actually building the building on the screen ahead of building it out in the field. So, it is relatively easy to sort out errors by way of moving the mouse around and moving something, as opposed to moving it after you build it. So, you're building the building twice basically. I think that is going to have a huge effect on making things more efficient, lowering cost, and eliminating a source of changes. I think that is a factor. I also think, I'm not an expert in it, but I think the concept of lean construction and lean operations is starting to get a little bit of momentum in the industry. In terms of how a project is planned, just-in-time deliveries, and things like that, which is a little bit different than the typical model of identifying the things that float in a Critical Path Method (CPM) schedule, I think there are going to be some changes along those lines. The other one is the integration of the building information model and the project schedule in terms of a way to really show how a project will go together.

Early Applications of Construction Management

Several well-known buildings in the United States were executed in the 1960s employing CMs.

In the early days, most of these projects employed a system that would today be described as at-risk construction management. General contractors sought to develop a way to meet their clients' need for expert advice in preconstruction stages of a project. The system that evolved as a result combined consultant services in the early stages with a traditional contracting structure once construction was under way.

Madison Square Garden in New York City. The historic arena was among the early applications of professional CM, delivered by Tishman Construction Company. PHOTO COURTESY OF TISHMAN CONSTRUCTION.

One of the most prominent of these early projects was Madison Square Garden, a sports arena built astride an active railroad terminal and its approaches. This is the fourth arena in New York City to go by the name Madison Square Garden, and is one of the world's best-known sports and concert facilities. It also played an unintended role in the history of New York: The destruction in 1964 of the old Penn Central Railroad terminal, with its Beaux Arts façade, created such a storm of protest that one result was the creation of the New York City Landmarks Preservation Commission.

The "new" Madison Square Garden opened in 1968, as part of a multipurpose complex called Pennsylvania Plaza, which houses retail shops, restaurants, and a large office building in addition to the continuing railroad operations below street level.

Historic skyscrapers executed using CM to address all the complexities of the entire project cycle. PHOTOS COURTESY OF TISHMAN CONSTRUCTION.

Tishman Construction Company cites both the John Hancock Center in Chicago and the World Trade Center in New York, among the tallest buildings in the world, as examples of the company's early construction management achievements.

The World Trade Center, destroyed in the attacks of September 11, 2001, was originally the concept of David and Nelson Rockefeller, heirs of one of the world's great family fortunes. At the time, David Rockefeller was chairman of Chase Manhattan Bank, and Nelson was governor of New York. The site of the huge complex was previously an industrial area, particularly noted for a large number of electronics stores, which led to its being called "radio row."

Once again, the proposal to acquire this property, close five streets, and demolish 164 structures in a 16-block area aroused significant public indignation. The need to address public stakeholders while simultaneously innovating in engineering and design created a unique set of challenges. Moreover, the two skyscraper towers were merely the anchors of a large multibuilding complex.

The Importance of Being a Good Listener

J.T. (TOMMY) THOMAS, JR., CCM, FCMAA

What advice would you give to someone who wanted to become a construction manager?

› A person should be seeking out and interacting with the full aspects of construction, understanding every bit of specifications, plans, dealing with contractors, and also specializing in an area where they have experience. You don't just come out of school and become a construction manager. There is a lot of academia that teaches it that way, but it doesn't necessarily work. You've got to have some experience at being, let's say kicked around, being involved in the court cases, understanding what the importance is of the recordkeeping process, and then from there you can start performing as a construction manager.

What's the most important trait for a construction manager?

› The most important thing is to listen to people and understand and not necessarily take sides but listen to everyone, because as a construction manager, you are the monitor for the owner or you are working directly for the owner as a construction manager. It's important that you listen to all sides and you put together your best judgment, leaving the owner to make final judgment, but you give them the best advice that you can.

What is the greatest challenge facing a construction manager?

› Dealing with the various personalities that are involved. You have to—wherever you can—divorce yourself from the financial side, not turning a completely blind eye to it, but trying to make the decision the best you can.

The John Hancock Center in Chicago, built between 1965 and 1969, is a multiuse building that holds offices, restaurants, parking, and residential apartments.

In the 1970s, the U.S. Department of Health, Education and Welfare (later renamed the Department of Health and Human Services) employed the combination of construction management with guaranteed maximum price contracts for a variety of hospital, school, and laboratory projects built with federal funds.

Some of these examples, involving comprehensive management of multiple projects for a single owner over an extended period of time, were actually early expressions of the discipline known today as program management.

In time, the value of professional, consultant CM services in preconstruction became well accepted, and other professionals began to realize they could offer these services separately from actual construction—without being exposed to the numerous risks that came with construction. Architects and engineers began to view construction management as a value-added service they could provide to their design clients. This was the root from which today's "agency CM" grew.

A key step in this evolution came in 1970, when the General Services Administration (GSA), which administers all construction work for the U.S. federal government, launched a study to reevaluate its practices. The resulting report recommended that GSA begin using phased construction in conjunction with construction management. By the following year, GSA had begun to award CM assignments, and it completed its first projects using CM by 1975. In 1976, the 200th anniversary of the Declaration of Independence, the National Air and Space Museum opened on the National Mall in Washington, DC, as a high-visibility project executed using CM.

The museum houses many of the icons of flight, including the original 1903 Wright Flyer, Charles Lindbergh's Spirit of St. Louis, Chuck Yeager's Bell X-1, John Glenn's Friendship 7 spacecraft, the Apollo 11 command module, and a lunar rock sample that visitors can touch. Since opening, the building on the Mall has been the most visited museum facility in the world, attracting on average more than nine million people annually.

TWO FLAVORS OF CONSTRUCTION MANAGEMENT: AGENCY VS. AT-RISK

WHAT IS AGENCY CM?

Agency CM is a fee-based service in which the CM is responsible exclusively to the owner and acts in the owner's interests at every stage of the project. The CM offers advice, uncolored by any conflicting interest, on such crucial matters as:

- Optimum use of available funds
- Control of the scope of the work
- Project scheduling
- Optimum use of design and construction firms' skills and talents
- Avoidance of delays, changes, and disputes
- Enhancing project design and construction quality
- Optimum flexibility in contracting and procurement

Comprehensive management of every stage of the project, beginning with the original concept and project definition, yields the greatest possible benefit to owners from construction management.

WHAT IS AT-RISK CM?

At-risk CM is a delivery method that entails a commitment by the CM to deliver the project within a guaranteed maximum price (GMP). The CM acts as consultant to the owner in the development and design phases, but as the equivalent of a general contractor during the construction phase. When a CM is bound to a GMP, the most fundamental character of the relationship changes. In addition to acting in the owner's interest, the CM also protects him/herself.

Tishman Construction handled CM for the accelerated delivery of the Walt Disney Company's Experimental Prototype Community of Tomorrow (EPCOT) at World Disney World in Florida. PHOTO COURTESY OF TISHMAN CONSTRUCTION.

United States Job Outlook

According to the U.S. Department of Labor, jobs for CMs in the United States are expected to grow faster than the national average.

> Employment of construction managers is projected to increase by 17 percent during the 2008–18 decade, faster than average for all occupations. Construction managers will be needed as the level and variety of construction activity expands, but at a slower rate than in the past. Modest population and business growth will result in new and renovated construction of residential dwellings, office buildings, retail outlets, hospitals, schools, restaurants, and other structures that require construction managers. A growing emphasis on making buildings more energy efficient should create additional jobs for construction managers involved in retrofitting buildings. In addition, the need to replace portions of the U.S. infrastructure, such as roads, bridges, and water and sewer pipes, along with the need to increase energy supply lines, will further increase demand for construction managers. (*Occupational Outlook Handbook,* 2010–11 Edition, U.S. Bureau of Labor Statistics.)

WHAT DO THE LETTERS MEAN?

As with other professions, CMs often have letters after their names that indicate licenses and other qualifications.

Here's what they mean:

CCM: The Certified Construction Manager is someone who has voluntarily met the prescribed criteria of the CCM program with regard to formal education, field experience, and demonstrated capability and understanding of the CM body of knowledge. The program is administered by the Construction Manager Certification Institute (CMCI), an affiliate of CMAA.

CMIT: Construction Manager in Training, designation awarded by CMAA to undergraduate students and early-career professionals who have passed its examination. CMITs work with experienced mentors to chart out a career path with the goal of earning the CCM in the future.

FCMAA: Fellow, CMAA College of Fellows. The Fellows designation is one of CMAA's highest honors, conferred upon industry leaders who have made significant contributions to their organizations, the industry, and their profession. The goals of the College of Fellows are: to represent a diverse community of thought leaders who lend their knowledge and insight to the strategic issues facing the industry and profession; to identify and develop future leaders; and to take an active role in CMAA.

FAIA: Fellow of the American Institute of Architects (FAIA). Fellowship is an honor bestowed by the American Institute of Architects on architects who have made outstanding contributions to the profession through design excellence or contributions in the field of architectural education, or have otherwise advanced the profession.

LEED: The Green Building Certification Institute's LEED (Leadership in Energy and Environmental Design) Professional Credentials demonstrate current knowledge of green building technologies, best practices, and the rapidly evolving LEED Rating Systems. LEED is administered by the U.S. Green Building Council. There are three tiers:

Tier 1: LEED Green Associate
Tier 2: LEED AP (Accredited Professional) (with specialty)
Tier 3: LEED Fellow

PE: A Professional Engineer is an engineer who is registered or licensed within certain jurisdictions to offer professional services directly to the public. In the United States, registration or licensure of Professional Engineers is performed by the individual states.

The increasing complexity of construction projects requires specialized management-level personnel within the construction industry. Sophisticated technology; the proliferation of laws setting standards for buildings and construction materials, worker safety, energy efficiency, and environmental protection; and the potential for litigation have complicated the construction process. In addition, advances in building materials, technology, and construction methods require continuous learning and expertise.

Prospects should be best for people who have a bachelor's or higher degree in construction science, construction management, or civil engineering, plus practical work experience in construction. A strong background in building technology is beneficial as well. CMs also will have many opportunities to start their own firms.

In addition to job openings arising from employment growth, many openings should result annually from the need to replace workers who transfer to other occupations or leave the labor force for other reasons. A number of seasoned managers are expected to retire over the next decade, resulting in a number of job openings.

CM JOB TASKS: MEAN FREQUENCY AND IMPORTANCE RATINGS

		Frequency			Importance		
		N	Mean	Std Dev	N	Mean	Std Dev
I.	**PROJECT MANAGEMENT PLANNING**	**527**	**1.65**	**0.43**	**527**	**2.48**	**0.37**
A.	Develop and manage the Construction Management Plan with measurable goals and objectives	526	1.98	0.86	525	2.58	0.63
B.	Organize and lead project team by implementing project controls, defining roles and responsibilities and developing communication protocols	526	2.22	0.83	524	2.77	0.47
C.	Identify unique site conditions and their impact on construction sequencing and operations	526	2.38	0.79	525	2.68	0.53
D.	Develop contract administration and documentation procedures	519	1.71	0.79	524	2.62	0.56
E.	Identify elements of project design and construction likely to give rise to disputes and claims	525	2.34	0.74	524	2.71	0.51
F.	Develop strategies and procedures to avoid disputes and claims.	521	1.94	0.80	521	2.61	0.57
G.	Create Project Procedures Manual	526	1.10	0.55	522	2.22	0.75
H.	Prepare information required to support owner's funding requirements	524	1.65	0.76	522	2.43	0.74
I.	Determine particular skills and experience needed for the project	524	1.62	0.75	523	2.30	0.70
J.	Help owner pre-qualify designers, develop and manage a selection process	520	1.08	0.72	517	2.23	0.81
K.	Coordinate designer's duties with CM's and owners in contract language	521	1.27	0.84	517	2.36	0.76
L.	Develop trade contractor's scope of work for contract agreements	522	1.22	0.90	509	2.34	0.92
M.	Assess and evaluate bidder qualifications relative to project requirements	523	1.36	0.71	523	2.44	0.69
N.	Define expected outcomes of the information management system to provide timely flow of information to the project team	522	1.58	0.82	518	2.28	0.76
O.	Define interactions and relationships among project development team	525	1.60	0.85	520	2.24	0.76
P.	Define responsibilities and management structure of project management team	521	1.38	0.71	525	2.87	0.40

II. COST MANAGEMENT	527	1.38	0.49	526	2.24	0.48
A. Interpret and integrate conceptual budgets provided by the owner and assess impacts on the project cost	525	1.65	0.81	517	2.57	0.62
B. Develop project budget taking into consideration project and owner objectives, cost constraints, and procurement strategies	522	1.49	0.76	519	2.65	0.60
C. Define purpose and objectives of the cost management system utilizing available resources to develop, coordinate and implement the system	520	1.20	0.75	510	2.19	0.78
D. Determine need for, and utilization of project feasibility studies	520	0.90	0.73	503	1.85	0.86
E. Monitor cost as the design is developed	520	1.85	0.82	515	2.62	0.64
F. Develop cash flow schedule based on project progress schedule	523	1.55	0.74	514	2.25	0.79
G. Determine impact of external economic factors on project cost and methods of project funding	524	1.19	0.75	509	1.96	0.83
H. Select estimating techniques appropriate to the project and its phases	520	1.25	0.79	516	1.97	0.81
I. Define project parameters at the conceptual design stage and address issues that may be associated with the conceptual estimate	518	1.31	0.82	509	2.23	0.79
J. Specify cost monitoring methods and Frequency of updating, cost tracking and reporting	520	1.43	0.77	509	2.32	0.74
K. Review design documents for conformance with scope and budget requirements	522	1.80	0.77	516	2.64	0.58
L. Develop criteria for constructability reviews and lead the effort involved	522	1.38	0.70	517	2.49	0.64
M. Lead design phase team meetings	521	1.61	1.12	513	2.07	0.91
N. Establish effective value analysis program	520	1.00	0.73	507	1.89	0.81
O. Specify, develop and implement an effective schedule of values for prompt and equitable payment requests evaluation	519	1.49	0.71	514	2.33	0.77
P. Specify, develop and implement an integrated cost-loaded schedule	515	1.24	0.81	504	2.12	0.91
Q. Specify, develop and implement an effective procedure for controlling, analyzing and evaluating cost of potential claims	517	1.54	0.80	510	2.40	0.69
R. Use man-hour/cost forecasting productivity studies, impact analysis, efficiency losses, etc.	516	1.11	0.94	500	1.69	0.93
S. Determine contracting procedures and types; advantages and constraints	517	1.13	0.69	508	2.15	0.76

III. TIME MANAGEMENT	523	1.78	0.48	523	2.43	0.44
A. Develop project scheduling requirements and systems	515	1.39	0.74	513	2.49	0.65
B. Develop project program schedule	518	1.36	0.72	513	2.45	0.69
C. Manage design phase schedule	515	1.66	0.98	506	2.31	0.77
D. Develop construction schedule	519	1.62	0.86	512	2.69	0.56
E. Review detailed short term schedules with contractor(s)	516	2.57	0.69	518	2.65	0.56
F. Implement and utilize schedule applications at the appropriate level for project participants	516	2.00	0.82	511	2.29	0.74
G. Manage selection and utilization of scheduling reports	518	1.68	0.79	512	2.01	0.75
H. Monitor project milestones and schedule calculations to determine relative percentages for work progress	519	2.13	0.62	516	2.54	0.63
I. Develop and manage a critical path schedule	518	1.98	0.84	515	2.67	0.59
J. Revise design schedule to reflect the changes that will occur and implement these revisions	519	1.69	0.88	511	2.30	0.76
K. Determine appropriate period of schedule recovery during the course of a project	517	1.83	0.77	514	2.43	0.70
L. Analyze concurrent delays, compensable and non-compensable delays	518	1.92	0.79	514	2.52	0.66
M. Define procedures for management of schedules to avoid risks and mitigate potential claims	519	1.51	0.79	512	2.39	0.72
N. Identify team responsibilities and lead team member interaction	516	1.88	0.87	513	2.40	0.70
O. Perform impact analysis on known or potential construction changes	516	2.00	0.79	512	2.45	0.65
P. Establish project milestones, durations of specific phases of the project	518	1.52	0.72	516	2.51	0.62
Q. Review acquisition plan and design documents to verify constructability within established performance periods	513	1.39	0.74	511	2.25	0.72

IV. QUALITY MANAGEMENT	518	1.65	0.54	518	2.36	0.49
A. Prepare a Quality Management Plan	514	1.00	0.58	509	2.32	0.79
B. Select, organize and lead the quality management team	511	1.39	0.95	503	2.19	0.81
C. Manage conformance of work to contract documents during the construction phase	516	2.59	0.72	515	2.74	0.52
D. Monitor risk management and implementation of safety plans	515	2.17	1.02	507	2.57	0.67
E. Evaluate capabilities of inspection, quality control personnel and service companies	510	1.84	0.90	511	2.33	0.68
F. Monitor effectiveness of the contractor's QA/QC plan	513	1.96	0.92	511	2.33	0.71
G. Determine materials that require testing and assess performance results relative to project quality objectives	511	1.81	1.01	507	2.33	0.72
H. Perform an operating review of the design on behalf of the Owner	512	1.10	0.85	495	2.03	0.83
I. Perform a maintainability review on behalf of the Owner	508	0.93	0.81	492	1.89	0.88
J. Review design documents for coordination between disciplines	511	1.64	0.86	505	2.51	0.66
K. Ensure adequacy of specification in design documents to achieve a desired level of quality	510	1.53	0.81	509	2.48	0.66
L. Ensure review comments are adequately addressed during the design phase	509	1.78	0.81	502	2.58	0.61

V.	CONTRACT ADMINISTRATION	520	1.50	0.45	520	2.33	0.48
A.	Determine appropriate project delivery techniques, and define roles and responsibilities of the parties for each technique	513	1.30	0.76	511	2.43	0.70
B.	Administer design processes and procedures, identify design team responsibilities and review requirements for different design types	516	1.17	0.90	493	2.12	0.82
C.	Organize and lead team member communication and interaction	510	2.29	0.89	507	2.52	0.66
D.	Develop a construction contract procurement plan and the contract methods	515	1.10	0.69	505	2.29	0.76
E.	Manage the bidding process including the procedures for bid evaluation and awards	512	1.40	0.87	502	2.47	0.73
F.	Develop scope of work for bid packages	511	1.40	0.86	505	2.55	0.68
G.	Help prepare contract documents	511	1.50	0.84	506	2.45	0.71
H.	Establish contractor pre-qualification procedures	511	1.11	0.71	504	2.22	0.78
I.	Implement procedures for dispute avoidance and team building	510	1.41	0.78	505	2.30	0.75
J.	Review bid performance bonds, letters of credit and other bid related documents	509	1.18	0.83	499	2.22	0.84
K.	Prepare communications procedures for all aspects of a project	512	1.41	0.78	512	2.41	0.70
L.	Manage construction phase of the contract	511	2.75	0.62	69	1.86	0.35
M.	Develop checklist for contract deliverables	510	1.67	0.82	509	2.37	0.69
N.	Monitor contractor compliance with contract requirements	512	2.72	0.58	515	2.78	0.46
O.	Establish process for contractor's procurement, expediting, delivery and installation	515	1.22	1.00	489	2.03	0.89
P.	Define partnering process and potential benefits to all parties	512	1.06	0.73	500	1.91	0.87
Q.	Develop and manage the plan for the documentation required during the acceptance and closeout process	513	1.49	0.76	513	2.49	0.64
R.	Define team member responsibilities and timing for inspections and corrections	512	1.85	0.87	515	2.41	0.66
S.	Establish appropriate retention procedures for contract, record and turnover document files	513	1.20	0.69	509	2.17	0.77
T.	Establish procedures for tracking closeout items	512	1.41	0.74	511	2.34	0.70
U.	Perform scheduling and documentation for operational training	511	1.23	0.80	499	2.14	0.77
V.	Develop requirements for occupancy and start-up	510	1.15	0.69	500	2.29	0.76
W.	Review As-builts for accuracy and completeness	509	1.52	0.83	505	2.34	0.73
X.	Ensure compliance with commission process	511	1.45	0.94	500	2.29	0.78

VI.	SAFETY MANAGEMENT	518	1.05	0.57	517	2.21	0.64
A.	Investigate liability imposed by statute and state-level safety legislation	512	1.03	0.91	487	2.21	0.86
B.	Determine existence of insurance coverage for safety-related claims	510	1.11	0.83	493	2.29	0.81
C.	Determine applicability of Workers' Compensation Laws	508	0.77	0.81	482	1.94	0.98
D.	Pre-qualify contractors by checking safety records, Experience Modification Rate (EMR) and OSHA 200 form	512	0.90	0.73	491	2.19	0.83
E.	Understand incident, Frequency, and severity rates for comparison with national averages	512	0.94	0.77	484	1.92	0.89
F.	Establish Project Safety Manual and review contractor jobsite safety programs	513	1.23	0.84	502	2.43	0.77
G.	Establish Project Emergency Plan detailing coordination of police, fire, rescue, and emergency medical services prior to beginning construction	512	1.07	0.70	499	2.53	0.71
H.	Establish risks associated with on-site safety and the respective roles and responsibilities of on-site employers	512	1.39	0.99	497	2.36	0.79
I	Coordinate safety program among contractors and create Project Safety Committee	511	1.15	1.05	486	2.16	0.93
J.	Determine contractor and CM safety training requirements	511	0.99	0.80	492	2.19	0.83
K.	Establish role of the Project Safety Committee	510	0.76	0.75	483	2.00	0.93
L.	Identify, assess and allocate risk	509	1.48	1.08	493	2.25	0.85
M.	Identify types and requirements for project insurance	514	0.82	0.68	491	2.15	0.85
N.	Identify alternative forms of dispute resolution and their potential benefits to all parties	509	1.02	0.69	497	2.12	0.78

VII.	CM PROFESSIONAL PRACTICE	517	1.79	0.52	517	2.50	0.47
A.	Operate as a leader in your practice	513	2.70	0.67	509	2.72	0.54
B.	Adhere to ethics of professional practice and encourage the same among your team	513	2.90	0.36	511	2.92	0.28
C.	Determine what delivery method would best fit your project	508	1.47	0.84	504	2.50	0.65
D.	Create and define the legal relationships between the Owner-CM	508	1.18	0.86	493	2.56	0.68
E.	Establish CM fee structures and benchmarks	508	1.14	0.84	500	2.43	0.75
F.	Enforce CM agreements for the appropriate delivery process	509	1.70	1.00	499	2.39	0.72
G.	Monitor adherence to legal employment discrimination and federal compensation laws	511	1.52	1.09	490	2.21	0.84
H.	Ensure team awareness of employee, environmental, discrimination and liability issues	515	1.71	0.98	505	2.30	0.77

Core Competencies (What You Need to Know)

Possibly the most comprehensive statement of core competencies for professional CMs emerged from a study CMAA performed in 2006. In this study, survey respondents were asked to rank 120 common project responsibilities according to both their importance in a CM's work and the frequency with which the respondents actually engaged in these activities. These responsibilities fell into seven broad areas:

- Project management planning
- Cost management
- Time management
- Quality management
- Contract administration
- Safety management
- CM professional practice

The *project management planning* arena was ranked as the most important of all project-specific functional areas by CMs responding to the survey. (The only category ranked more important was maintaining CM professional practice.)

Among project management functions, respondents gave the highest importance to defining the responsibilities and management structure of the project management team.

Ranking second in importance was "organizing and leading the project team by implementing project controls, defining roles and responsibilities and developing communication protocols." Third place in importance was assigned to identifying elements of project design and construction likely to give rise to disputes and claims.

Survey respondents reported that their most frequent jobs—functions they performed on a daily or near-daily basis—included prequalifying designers, developing and managing a selection process, creating project procedures manuals, and developing trade contractors' scope of work definitions for contract agreements.

The next most important general area was *time management*, in which CMs reported their most important function is to develop a construction schedule, followed by developing and managing a critical path schedule for the job and reviewing detailed short-term schedules with contractors.

CMs reported they performed these functions on at least a monthly basis, with the greatest frequency being reported for "developing project scheduling requirements and systems," "developing project program schedule," and "review acquisition plan and design documents to verify constructability within established performance periods."

Respondents to the CMAA survey also identified a number of *quality management* functions as central to their jobs. Chief among these, in terms of the frequency with which CMs perform specific tasks, is to "manage conformance of work to contract documents during the construction phase." Two other key functions tied for second in importance: "Monitor risk management and implementation of safety plans" and "ensure review comments are adequately addressed during the design phase."

CMs reported performing these functions on a monthly basis. Other jobs performed on a monthly or weekly basis included preparing a Quality Management Plan, selecting and leading a QM team, performing design reviews on behalf of the owner, and monitoring the effectiveness of contractors' QM/QC teams.

Contract administration is a central CM function, the survey found. CMs attached the highest importance to monitoring contractor compliance with contract requirements, developing scope-of-work documents for bid packages, and organizing and leading team interactions. The most frequently performed functions reported by CMs included defining the partnering process, developing a contract procurement plan and contracting methods, establishing contractor prequalification procedures, and developing requirements for occupancy and start-up.

Construction management also focuses significantly on project costs. In the field of *cost management*, the survey found CMs and owners defining the budget development process as their most important function. Next came reviewing design documents for conformity with budget and scope requirements, and monitoring costs as the design is developed.

Addressing *safety management*, CMs said the most important function they perform is to establish project emergency plans, including coordination of police, fire, rescue, and other emergency services. Establishing risks associated with on-site safety, defining responsibilities of on-site employers, and reviewing contractor safety programs were also seen as highly important.

CMs and owners responding to the survey ranked maintenance of *CM professional practice* as their most important ongoing concern, including adhering to ethical standards and providing true leadership to their firms.

In 2008–2009, a special committee of CMAA developed a revision of the *CM Standards of Practice* (*CMSOP*) to reflect changes in the industry over the years. The new 2010 edition of the *CMSOP* added three critical areas to those in which CMs would require important professional skills: risk management, program management, and sustainability. Each of these areas was addressed in detail in the 2010 *CMSOP*.

Developing the CM Approach

JOHN TISHMAN

Founder and former chairman, Tishman Construction Corporation

How did CM first arise as a specialty for your company?

› In the 1950s and 1960s, I was able to see and understand the problems that arose between owners and general contractors, perhaps more acutely than other developers or contractors, because as an owner/builder doing construction for our family's portfolio, I was able to avoid many common pitfalls. For example, Tishman Realty and Construction did not as often run into the problem of shifting designs in the midst of the construction phase or, on the construction side, of employing a GC that needed to cut corners in order to make a profit.

The process that I labeled Construction Management is essentially what we at Tishman Construction had been doing for Tishman Realty—managing the construction.

The first opportunity for managing came in supervising the linked construction of Madison Square Garden and Two Penn Plaza. For that project, we were paid a fee that was a small percentage of the total construction budget. Coming at this task from the point of view of owner-builders, we looked for ways to slim down the construction time, and to use the best systems, those that would lower later operating costs, because every nickel saved and week of construction that we did not have to do was reflected in our bottom line and in the bottom lines of our partners.

How did this role differ from general contracting?

› In a sense, we at Tishman Construction were performing the supervisory tasks that a general contractor would do, but we were doing these tasks from a position *on the same side of the table as the owners*, working with them instead of, as a GC must do, negotiating against them. This meant, for example, that we did not have to find ways to make a little extra money from the owner on Part A of the project as a result of having underestimated a cost in Part B so we would lose money on that . . . because we were not struggling to make a profit as a contracting entity; rather, we were managers, being paid a fee for our expertise and our supervisory services.

We became a construction entity that built for an owner as though we were part of the owner's team. In the Madison Square Garden project, the owners had, in effect, rented the Tishman Construction division, and my construction division colleagues and I functioned as their employees, although we did it from our own Tishman offices.

What came next?

› Right after the MSG project, I was able to successfully argue that our construction division should be allowed to look for other such fee-based opportunities to mange construction projects for others. I was authorized to take our owner/builder approach to Construction Management to school, so to speak, with projects for the University of Ohio, New York University, and the University of Illinois.

The third one was particularly significant for the history of CM, because it required us to convince the federal government, whose funds were being used to finance the university library, that CM was not only permissible in federal contracts but desirable.

The Perspective from Government

GEORGE LEA, JR., FCMAA, CCM

Program Manager

U.S. Army Corps of Engineers

What is your background?

› I graduated from Bucknell University as a civil engineer, and I wasn't sure I wanted to do engineering so I did a work study program during my last year and worked at an engineering office and, at the same time, went ahead and was actively being recruited by the Navy to fly airplanes. So I didn't know which one I was going to do, but what I did learn in my assignment was that I can do this engineering stuff any time, but I only have one chance to fly airplanes. I went and signed up for the Navy to fly for seven years, and I did that off the aircraft carrier, knowing that I would return to engineering again, and that is exactly what I did. After seven years as a pilot, I went back into engineering and I was working as a project manager, and I was figuring that I didn't have the technical background to really do the work. I had a technical background but not like my peers did who were already at the same level as me.

What did you do next?

› I figured I needed to get more knowledge. I went overseas and eventually convinced them to give me a director of engineering assignment because I had all the leadership skills. I worked to get my PE at the same time, while managing an office and progressing my career in the technical side of engineering, and then moved over to construction, with the Navy. I was the director of a large billion-dollar relocation and construction program and improvement initiative. I ran everything underneath me as the deputy program manager: engineering, construction, design, and real estate planning project. I did that for about seven or eight years, then I applied to the Corps of Engineers for a job to be the chief engineer program manager on the Pentagon renovation. So I was at the Pentagon renovation for two years, maybe just a tad longer, and then I moved over to the Baltimore district with the Corps to be the chief of construction. I was the chief of construction for six years and then moved down into headquarters here, and learned a little about the civil works program. I left the government for six months and was the chief engineer at DC Metro, and then they fired the general manager who brought me there, and I figured that wasn't a good place to be, so I went back with the Corps, and now I am the deputy of the military side of engineering and construction, in the military branch of engineering and construction.

How important is it that the government recognizes construction management as a profession?

› I think it is absolutely critical. Certain agencies are suffering because they can't hire the right people with these construction management degrees. If it is just the federal government, we'll get by, but this whole evolution is much, much bigger than that. It's about the federal government recognizing the profession, and as soon as we do that as a federal government, what happens is that we start requiring in our contracts that the person managing the job meet the same level of professionalism to which we hold ourselves.

How important is it to have people who are certified CMs?

› I think it is essential. Certification establishes a profession level. It should be the minimum standard to manage construction in today's environment. I have a dream that in the future we might be able to write in legislation and say: "Here's two billion dollars to do infrastructure projects, and a construction manager must be on the job, a certified construction manager." Utah was the first state to require licensing of construction managers through the certification program. I hope other states will follow their lead.

What is different about being a CM in the federal government, as opposed to private industry?

› The big thing is risk. It goes very deep into the process and management of construction management. The understanding of risk is absolutely critical, and every element of the job, everything you do or touch, should have to be balanced to have a successful job. In the federal government, we put a lot of risk on the contractors. It's in our contracts.

A Demanding Career

Construction management is a demanding profession. Many decisions have to be made on a daily basis. There can also be considerable travel if the construction site is not close to the main office, or when there is responsibility for two or more sites. If you're going to manage overseas projects, you will have to relocate to the host country.

You are often on call 24 hours a day, dealing with severe weather, accidents, work stoppages, or other emergencies at the job site. Quite often, you will work more than the standard 40-hour week, especially if a project is behind schedule. Indeed, some projects may continue work around the clock.

Unlike some construction jobs, being a CM is not inherently dangerous, but injuries can occur, and CMs must be careful while visiting project sites.

The Construction Management Association of America (CMAA)

The Construction Management Association of America (CMAA) is North America's only organization dedicated exclusively to the interests of professional construction and program management.

It was established in 1982, and its website is www.cmaanet.org.

CMAA membership topped 8,000 in 2011, including individual CM/PM practitioners, corporate members, and construction owners in both public and private sectors, along with academic and associate members. CMAA has 28 regional chapters.

CMAA presents two national gatherings annually, which offer extensive educational programming as well as opportunities to meet and network with colleagues.

The Construction Manager Certification Institute (CMCI), a subsidiary of CMAA, administers the Certified Construction Manager (CCM) program, which has been accredited by the American National Standards Institute (ANSI) under the terms of the International Organization for Standardization's ISO 14024, governing the management of personnel certification programs.

The CMAA Foundation is active in career promotion, research, and scholarships for college-level studies in CM and related fields.

History of CMAA

In 1981, some three dozen people, including architects, engineers, and constructors, met in Indianapolis, Indiana, and formed a steering committee to begin the work of organizing a new association for professional CMs. The new national association's founding board of directors included representatives of CM firms such as Heery Program Management, Parsons Brinckerhoff Construction Services, Hill International, and CH2M HILL, along with architectural and engineering concerns such as O'Brien-Kreitzberg & Associates. The association's first national meeting occurred the following year in Denver, Colorado.

The new association declared that it had two key missions: first, to foster understanding of CM among owners, potential clients, and other professional groups; and second, to develop CM as a profession, including a code of ethics, standardization of practices, and other factors.

Jerry Klauder, a CM veteran and future president of the association, recalled that in the early 1980s, "as an emerging profession, Construction Management had existed in an atmosphere of opposition and a cloud of controversy for over a decade. It faced many problems. . . . [and] the nature and magnitude of the problems suggested that it would be difficult, if not impossible, to solve them as individuals."

Since professionalism was critical to the image of CM and to building new opportunities for its practitioners, the association would assume the task of promoting professional development through standards of practice, education, and communication.

Most urgently, this professional development mandate required that the new association take the lead in defining the content of the profession—what individuals needed to know, and what skills they needed, to perform at a high level. CMAA approached these challenges by working to develop standards of practice for the profession and, later, to measure and define the profession according to the specific job functions its leading practitioners were actually performing for their clients.

CMAA's National Conference & Trade Show provides an annual opportunity for networking and continuing education. CMAA

Shortly after CMAA was organized, a Standards of Practice Committee was formed to begin the job of codifying what construction management, properly performed, really meant. Creating the first draft of the *CM Standards of Practice* (*CMSOP*) took more than two years.

Published in 1986, the new *CM Standards of Practice* drew possibly predictable reactions. *Building Design and Construction* ran an article in which owners expressed the view that *CMSOP* did not give the CM enough authority and involvement, while architects opined that the document gave CMs too much.

Very soon after publication of the initial *CMSOP* document, CMAA undertook an extensive revision, which was published in summer 1988. For the most part, the changes in the second edition were organizational. For instance, the *CMSOP* Committee reported to the Board of Directors:

"The new document has been substantially reorganized in terms of where, within the original phase and functional structure, each of the tasks is defined.... The original, which in effect was a composite effort of a number of contributing authors, has been edited for consistency of format and style. Both the reorganization and a new, paged table of contents will improve the usefulness of the document as a reference."

With the *CM Standards of Practice* gaining acceptance, CMAA was in a position to certify CMs, and in 1986 it began work to develop a system for certification.

CMAA MISSION & VISION

"The Mission of CMAA is to promote and enhance leadership, professionalism, and excellence in managing the development and construction of projects and programs.

"The Vision of CMAA is to be the recognized authority in managing the development and construction of projects and programs.

"CMAA is leading the growth and acceptance of construction management as a professional discipline that can add significant value to the entire construction process, from conception to ongoing operation.

"All parties to a project are vitally interested in excellence of execution—including rapid completion, high fidelity to specifications, conscientious cost control, and optimum use of all resources. Professional construction management delivers these values" (http://cmaanet.org/cmaa-mission-and-vision).

Ed Rendell, then governor of Pennsylvania and a founding co-chair of Building America's Future, addresses the CMAA National Conference on infrastructure investment priorities.

Charles Thornton, founder of the ACE Mentor Program, explores future CM manpower needs at the CMAA National Conference & Trade Show.

Providing Leadership during a Tough Government Project

BOB HIXON, JR., CCM, FCMAA, PE

Director of Federal Government Services, McDonough Bolyard Peck

Tell me about your work in the federal sector.

› I was with the federal government for 37 years, but I am now in the private sector and have been for a little over four years—but a lot of what I have been doing in the private sector is still doing government work.

You were involved in the Capitol Visitor's Center, right?

› I went to the Capitol Visitor's Center as a consultant for GSA, which is where I worked for 34 years. The GSA commissioner was asked by the Architect of the Capitol for some assistance in that project because the Architect of the Capitol did not often do very large projects. GSA of course was doing courthouses and federal buildings all the time. At the time, I was the Director of Construction and Project Management for the GSA in the office of the chief architect. Bob Peck, who was the commissioner at the time, had me go over with him to the Capitol and sit through a briefing about their planned project, and then later on they were asking for some assistance with some issues they were having with the construction contractor, and I provided some comments on that.

They had added a lot of work after the September 11th bombing, so as a consequence of that there was a lot of new work, which was essentially building out 170,000sq/ft of the 580,000sq/ft project. That had an impact, of course, and the contractors knew the impact, but apparently the impact was being downplayed, like, "Oh, we can take care of that, no problem. . . ."

At that point, things were delayed, costs were going up dramatically because of this added work, and there was a lot of discontent about Congress. They are supposed to provide oversight over the rest of the government, and their own project was running out of control. I was asked to take over the project, and at the time I had been running GSA's construction and project management for four years. And, you know, when you get in and do something like that for four years, when you start it up and get it going, after a while you are not doing anything different. You are just kind of tweaking, and I thought, "Well that might be interesting, to take a challenge like a construction project again." I had previously done the Reagan Building (Ronald Reagan Building and International Trade Center) here in Washington, which was GSA's largest building at the time, so I said, "OK, I will do that."

A lot of what I did in the beginning was just trying to keep everybody from attacking the team so that they could stay focused, because if everybody is being attacked on the quality of the design or the quality of the construction or the quality of the CM, they stop focusing on what the project needs and start focusing on what they have to do to keep from getting hurt. So my role in life predominantly was to raise the umbrella over the team so they could stay focused, and when I asked them to do something they could do it and not worry about being attacked over it, because if anyone was going to be attacked over it, it would be me.

That was the first part; the other part was that the CM on the project was trying to manage the job, but yet the Architect of the Capitol people on staff were making their own decisions, and so it kind of rendered the CM powerless. I had to deal with that issue as well as preparing for congressional hearings [about the cost delays]. I met every Monday afternoon with the congressional staff, some 18 staffers, including the Secretary of the Senate and Clerk of the House, to talk about the status of the project and what was going on. Their role in life was to provide oversight on how the project was going and make sure it went well. So there were all of those influences, and we had to work though dealing with all of those, and it also affected how we could get this thing done while everybody peered over our shoulder.

Was your most important role to be running interference for your people?

› I think the most important role I had to play was to provide leadership, and the idea on providing leadership was to be the front for the team, to express what we were doing effectively and succinctly to the external stakeholders. We had to express to them what we were doing, why we were doing it, why it made good sense, and to protect the team from being individually attacked from outside the team. I think leadership was my primary focus.

What was it that attracted you to become a CM?

› In the very beginning, when I was just graduating out of college, you did job interviews and you really didn't know what you were getting into, so I just took a job with GSA. I started out as an inspector on a job site, but what kept me in it instead of going into design firms or something like that was I enjoyed managing projects, trying to get things done effectively, dealing with various groups of people, rather than being a designer who would sit at a board and be creative; that wasn't really my strong suit. I was better at working with people to try and figure out ways to get things done and get them done right. The government handed me as much rope as I wanted to take, and I ended managing projects at a much younger age than you would if you were on the outside.

What do you think are the most important skills a person needs to be a CM?

› You need to be able to work well with people, effectively, which means you have to listen, rather than always talking. You have to understand everybody's position so that you can determine the most effective way to accomplish things. You are motivating people to work with you to effectively complete a project, and individually those people have different things that motivate them or different goals. You have to find a common thread with the team so everyone stays focused on getting the project done successfully, rather than making the most money they can.

What would you tell a student, someone who is interested in being a construction manager; what is the least satisfying part of the job, and what is the most satisfying part?

› The least satisfying, I think, is if you are dealing with people who typically know a lot more then you do about construction. It can be intimidating because you are supposed to be orchestrating this parade for success, and yet you have people who, especially in the beginning, know a whole lot more about it than you do. So it becomes frustrating because the team you get is the team you get, and you are forced to try and work with people, and because of your younger age you may have some difficulty. On the other hand, the most satisfying thing is realizing that most people really want to do a good job; they're not out to screw the project, and they're not out to take advantage of everybody. They really want to do a good job quickly, efficiently, so they make a reasonable return on their money and move on to the next one.

What Does CMAA Stand For?

"The professional construction management services concept has grown out of a universal recognition throughout the design and construction community that quality-focused, cost-effective, dispute- and injury-free project delivery does not occur without a deliberate commitment and effort to manage the project delivery process. The cost and complexity of today's capital projects, the importance of time, and the need to deal with unanticipated events and unforeseen conditions all argue the need for an integrated and managed approach to planning, design, and construction of the built environment.

"Owners are faced with an array of project delivery options, each of which involves costs and benefits matched against the needs of the owner and the specific project. The task of management is to ensure that the value of those various project delivery methodologies is realized in achieving the owner's objectives.

"Professional construction management services are selected on the basis of qualifications and experience matched to the needs of the owner and the project and compensated on the basis of a negotiated fee for the scope of services rendered. This approach results in a cost-effective, dedicated representation of the owner's interests free from potential conflicts with regard to the cost of the project" (http://cmaanet.org/cmaa-mission-and-vision).

Skills

One of the characteristics that makes construction management so compelling to many people is that it combines a lot of different skills rarely found in other jobs. According to the Department of Labor (www.onetonline.org/link/summary/11–9021.00), skills needed for being a CM include:

Time Management—Managing one's own time and the time of others.

Active Listening—Giving full attention to what other people are saying, taking time to understand the points being made, asking questions as appropriate, and not interrupting at inappropriate times.

Critical Thinking—Using logic and reasoning to identify the strengths and weaknesses of alternative solutions, conclusions, or approaches to problems.

Management of Personnel Resources—Motivating, developing, and directing people as they work, identifying the best people for the job.

Speaking—Talking to others to convey information effectively.

Complex Problem Solving—Identifying complex problems and reviewing related information to develop and evaluate options and implement solutions.

Coordination—Adjusting actions in relation to others' actions.

Monitoring—Monitoring/Assessing performance of yourself, other individuals, or organizations to make improvements or take corrective action.

Negotiation—Bringing others together and trying to reconcile differences.

Active Learning—Understanding the implications of new information for both current and future problem-solving and decision-making.

What Does the Federal Government Think of Construction Management?

One of the important issues in the construction management community is the acceptance by the U.S. federal government of the profession. Should federal projects require or recommend that CMs be hired for their jobs? Indeed, should construction management receive a unique federal job definition, as do other professions or jobs?

In 2009–2011, a national working group studied the subject. The group included representatives from public and private owners, federal and nonfederal agencies, large and small businesses, trade associations, accrediting bodies, certification institutes, and academic institutions. In addition to the Army, federal agencies represented included the U.S. Navy Facilities Command (NAVFAC), the U.S. Air Force, the GSA, and the Veterans Administration (VA).

The working group reached some conclusions that will be vital in shaping the future of CM as it is practiced in the federal government environment. Among those conclusions:

1. The construction industry has changed profoundly in the decades since the Office of Personnel Management's (OPM) job series and position descriptions were developed. Construction projects and programs are being executed today in ways that are fundamentally different from the processes of the past.
2. The profession of construction management has emerged and gained acceptance throughout the design and construction industry as a result of these changes. Industry-wide trends, sustained over many years, have demanded that construction owners in private industry approach their projects and programs in new ways; the federal government must follow suit in order to realize the full benefit of modern concepts and methods.
3. Construction management education is delivered by institutions accredited by the ACCE as well as the ABET. Although private industry recognizes these two accreditations equally, federal job definitions at this time do not.

The working group's arguments give a comprehensive view of the construction management profession. These are worth reading because they offer excellent arguments, in general, for the promulgation of the profession of construction management.

The group noted:

1. *The construction industry has changed profoundly.*

For decades, the government worked in the established bid-design-bid-build system, which placed clear demarcations between the design and construction processes.

In defining jobs related to construction, the government's guiding principle has been that construction projects entail (1) a project engineering function, i.e., design, and (2) a construction engineering function. In the past, it was realistic to maintain a distinction between these functions; each position description accurately described what the holder of that job was expected to do.

Several major trends have swept this environment away. First, construction projects have become immeasurably more complex. Second, dramatic inflation of construction costs, coupled with persistently constrained public financial resources, has created great pressure for more effective ways of executing projects. Third, a series of innovative new project delivery methods have come into widespread use, many of which derive their success and effectiveness from blending, rather than separating, functions.

The professions of architecture and engineering, in themselves, are more complex and challenging today than ever before. The American Society of Civil Engineers (ASCE) maintains that a four-year college degree is no longer sufficient preparation for a career as a professional engineer; instead, a minimum of six years' education is now required, including both an undergraduate degree and a specialized postgraduate degree.

Architects today must contend with the fact that the design function itself has changed and is more widely disseminated through the construction team than previously. In a 2009 white paper on Integrated Project Delivery, the College of Fellows of the Construction Management Association of America noted that "Industrialization and a competitive environment have driven manufacturers to develop more and more sophisticated building products. The result is that architects and engineers include more and more in their design that they did not design and do not fully understand.... The AE's job has changed: It is to evaluate and integrate systems and products designed by others."

The increasingly collaborative nature of today's construction industry—powerfully enabled by new technologies—argues for a new approach to defining and categorizing the skills that contribute to success.

High inflation, particularly in the construction industry, has also created a demand for faster and more efficient project execution. Such techniques as "fast track," bridging, design-build, and others have emerged in response. All of these project delivery methods have in common that:

- They require multidisciplinary teams.
- They blur the traditional lines between functions.
- They are ideally led by individuals whose knowledge, skills, and abilities are those of a professional construction manager, rather than an architect or an engineer.

2. Construction management has emerged as a profession.

It has long been recognized throughout the construction industry that construction management is properly performed by an individual who has obtained a four-year degree from an appropriately accredited educational institution. Firms public and private, large and small, hire graduates of four-year university programs to manage construction. There remain some CMs in the industry who have learned their jobs through a combination of education and experience; these individuals may not have four-year degrees. This is not the industry's preferred standard, however. It is expected that the normal replacement of retirees by industry newcomers will result in an all-degreed professional cadre within the foreseeable future. In general, no federal agency today will hire a non-degreed individual to manage a construction project of other than minor scope and significance.

The academic community, beginning as long ago as the 1950s, has developed curricula and course content relating to construction management. These curricula were refined with the goal of educating an effective professional CM. The Curricula Development Subcommittee of AGC's Construction Education Committee identified, as long ago as 1967, the essential qualifications of a CM.

A CM, the subcommittee said, should have an education that would contribute to (l) the human understanding to work with all types of people, (2) the discipline to think and reason logically, (3) the technical ability to visualize and solve practical construction problems, (4) the managerial knowledge to make sound decisions and communicate them on a prudent economic basis, (5) the facility to communicate these decisions clearly, and (6) the professional stature to provide dynamic leadership in the construction industry and the community of which he or she is a part.

A broad consensus has emerged concerning the correct content of this education. Universities and colleges nationwide offer degree programs with very similar course sequences, core requirements, elective courses, and other features. There is a high degree of consistency in the education being provided to future CMs.

For its part, industry has responded to the need for specialization in construction management by developing broadly accepted standards of practice, a body of knowledge, a code of ethics, career-long continuing education programs based on these SOPs, consistent outreach to and collaboration with educational institutions, and personnel certification programs.

A View from Both Sides

JUDITH KUNOFF, CCM, AIA, LEED AP

Chief Architect

MTA New York City Transit Authority

What are you doing now?

› I'm at the New York City Transit Authority. I'm the chief architect. I'm not doing CM at the moment, but I surely am going out to construction sites and yelling at CMs, knowing full well what it is I expect them to do, since I was doing it.

You were trained as an architect. Why move to being a construction manager?

› I thought that CMs were better compensated for their efforts. I also thought that an architect, or any designer, should spend a stint out in the field. I thought they would become better designers. It lets you see how what you're putting on paper actually translates in the field. So, I set out to find a CM job, but of course I had no CM experience. As it happened, I had a relative who was looking for someone to design and oversee construction of a biotech company. It wasn't pharmaceutical, and it wasn't exactly manufacturing, it was something in between. I was their architect/CM, and it was great fun. I got to design and then take it out to the field.

Then, I had the opportunity to work for Skidmore, Owings & Merrill as a designer again. I got my master's degree in Israel, the Israel Institute of Technology. I lived in Israel for around five years.

Later I went to Parsons Brinckerhoff as a CM. That worked out really well. I worked for PB for 10 years as a CM. I was very successful there; I really developed and grew as a CM.

As a CM, what do you think, personally, for you is the greatest challenge in that job?

› The greatest challenge is to make sure that the owner, your client, respects you, sees you, trusts you, believes in you, and knows that you're out there to do everything for them. But, by the same token, [make sure] that the contractor respects you as well and sees that working with you is going to yield the best product.

Is any of this related to gender? It's obviously overwhelmingly male dominated, at least in the construction side of things. Is that ever an issue?

› I expected it to be an issue. I think probably people raised the red flag that this would be an issue for all of the reasons that you are thinking. What wound up happening was that, because I know my stuff and I know, at least to a certain extent, what I don't know, I have no fear of asking questions; I am comfortable asking questions. They were very respectful in the end. I think they were more uncomfortable in the beginning, now it's probably more commonplace to have women than it was 15 years ago.

What is the most satisfying part of being a CM?

› Seeing the project come to fruition. That's a no-brainer.

What is the least satisfying thing?

› Rework. Rework can happen for all kinds of reasons. It is very frustrating for all parties.

Now you have seen it from both sides, you've been a designer and you've been a CM. Which one do you like better?

› I like them both. I understand that all designers won't, and not all designers are going to be successful in both capacities. So, to tell you that I think all designers should have the experience is just a pipe dream. I think everybody benefits from it. I look around at the guys who work with me, whom I think of as successful. You don't necessarily have to be a CM, but at least to spend a real good stint out in the field makes all the difference. And, it is much easier to have a conversation then with a contractor out in the field, when you're frustrated with what he's done with your design, to be able to talk to him about how to build it. And most designers, if they haven't had the field experience, they really only know how to do it on the board, and nowadays in the computer room.

CM Education

Professional CMs today generally have degrees from programs accredited by one of two leading organizations: ABET, which accredits programs granting the traditional family of engineering degrees, and ACCE, which accredits programs offering degrees in construction science and construction management. (An in-depth discussion of education and the CM is in Chapter 2.)

At present, ABET has accredited 23 programs that offer the Bachelor of Science degree in construction engineering. ACCE has accredited 62 programs that lead to a BS degree in construction management. The academic community, then, has clearly recognized construction management as a professional discipline and has provided ample opportunity for interested individuals to obtain appropriate educational preparation.

Federal agencies tend to one of two viewpoints concerning which type of degree is appropriate for a CM.

The Naval Facilities Engineering Command (NAVFAC), the Corps of Engineers, and the U.S. Air Force are among the agencies that generally require an engineering-centered degree. NAVFAC views this as an important requirement in order for the CM to (1) maintain the role of the building official/authority, having jurisdiction and "responsible charge" (independent control and direction of the investigation or design of professional engineering work, or the direct engineering control of such projects), and to (2) maintain command-wide engineering expertise (for career progression and assistance to forward deployed military engineers), because project sites are often remote and may have limited availability of technical resources to the construction management team.

A new bridge to replace an older dam-top roadway at Hoover Dam in Nevada was one of the most spectacular infrastructure projects of recent years. CM by HDR, Inc. PHOTOS COURTESY OF HDR, PHOTOGRAPHER: KEITH PHILPOTT.

Other public or private firms and federal agencies such as GSA will almost never assume technical risk or liability of a project by allowing the CM to unilaterally make engineering decisions. These agencies generally express no preference concerning the type of degree the CM should have.

Due to the complex nature of medical facilities design and construction, the Veterans Administration recruits only individuals with engineering or architecture degrees to manage its major construction program.

Many agencies, and many authorities within the Corps of Engineers and throughout industry and government, will say that a business-, management-, and leadership-centered degree in construction sciences, as offered through the ACCE accreditation, best prepares the graduate for a career in construction management.

The working group concluded that these two types of degrees are accorded equal recognition throughout private industry when hiring CMs. Moreover, it is in the government's best interest for all agencies to be able to draw equally from both talent pools in meeting their construction management needs.

It is common, moreover, for individuals to approach the CM profession with degrees from other programs. The CMCI has identified a broad range of degrees that could potentially qualify their holders to become CCMs. These degrees include:

Construction Management

Electrical Engineering

Construction Science

Chemical Engineering

Civil Engineering

Architectural Engineering

Mechanical Engineering

Architecture

Broadening the types of educational backgrounds suitable for a professional CM will help both the government and the private sector access the largest possible pool of talented employees. This will be especially critical in the future. As older, experienced employees retire, their skills will need to be replaced, and new entry-level workers will be needed. A number of national observers have raised concerns about a growing shortage of engineering graduates from American universities—not only a decline in the number of grads, but an increase in the number being lured into other fields, such as finance, by the prospects of better pay and advancement opportunities.

2 The Education of a Construction Manager

UNLIKE SOME OTHER PROFESSIONS, medicine and law for example, construction management does not have a clear-cut academic path. Some people become CMs after logging many years of experience in the construction industry—the "school of hard knocks" route. However, in our current environment, more and more CMs are entering the field through academic training mixed with on-the-job and entry-level experience, such as college courses coupled with internships or summer jobs.

Even when colleges and universities offer specific construction management courses, their breadth varies from school to school, both on the undergraduate and graduate levels. Many colleges and universities offer four-year programs in construction management. Others round out formal education in architecture or engineering by allowing graduates to be hired as entry-level CMs, working as assistants to CMs, field engineers, schedulers, or cost estimators. Numerous colleges and universities offer a master's degree program in construction management or construction science. Master's degree recipients, especially those with work experience in construction, typically become CMs in very large construction firms or construction management companies.

As for two-year or junior colleges, many of these offer construction management or construction technology programs, and many individuals also attend training and educational programs sponsored by industry associations.

According to the U.S. Department of Labor, a bachelor's degree in construction science, construction management, building science, or civil engineering, plus work experience, is becoming the norm. However, years of experience, in addition to taking classes in the field or getting an associate's degree, can substitute for a bachelor's degree. Practical construction experience is very important for entering this profession, whether earned through an internship, a cooperative education program, a job in the construction trades, or another job in the industry. As mentioned, some people advance to construction management positions after having substantial experience as carpenters, masons, plumbers, or electricians, for example, or after having worked as construction supervisors or as owners of independent specialty contracting firms. However, as construction processes become increasingly complex, employers are placing more importance on specialized education after high school.

More than 100 colleges and universities offer bachelor's degree programs in construction science, building science, and construction engineering. These programs include courses in project control and development, site planning, design, construction methods, construction materials, value analysis, cost estimating, scheduling, contract administration, accounting, business and financial management, safety, building codes and standards, inspection procedures, engineering and architectural sciences, mathematics, statistics, and information technology. Graduates from four-year degree programs usually are hired as assistants to project managers, field engineers, schedulers, or cost estimators. An increasing number of graduates in related fields—engineering or architecture, for example—also enter construction management, often after acquiring substantial experience on construction projects.

CM education continues throughout a career, including timely professional gatherings that offer continuing education units, learning hours, and other credits for obtaining or maintaining credentials. CMAA

Education Is the Key to Moving Ahead in Construction Management

MARK HASSO, PhD, PE

CM Program Coordinator

Wentworth Institute of Technology

As far as education is concerned, what is your best advice to somebody already in college?

› They really need to be focused on why they want to be in this business, because I see a number of transfers, about 30 or 40 a year who come from other departments and other schools, and I do not encourage them to get in unless they are focused and they know what they are doing. Many of them would come in and say, "Somebody told me that this is a good field," and I would say that is not enough. You need to research and understand why you want to get into this business because you don't want to change your mind again. Switching is a problem; I have seen students who have switched one or three times until they got into construction.

This is a rewarding business if you know what you are doing, and the compensation is relatively higher than in other engineering disciplines. That's the financial side. On the practical side, you are always doing something different, so it is not monotonous, and the profession has been evolving over the past 20 or 30 years. It has become more exciting.

Students should look at what they want to do and what fields of interest in construction management they would want to get into. So, I would also expose them to the different specialties that they can get into, depending on their interest: They can go into the financial side, the estimating side, the scheduling side; they can go into the field; they can go into the project management side. There are so many ways that they can get into CM. It is not limited, and it is a much broader business than any other business, so that is quite attractive to many people for that reason.

And what about somebody who is already in the construction business?

› I treat tradespersons the same way. I will tell them that, as a tradesperson, your work is limited by your age because of the hard task you must accomplish on the job. You may not be getting more money at first than those who graduate, but in the long run the advantage of being in this field [CM] is that you will be able to continue working whether you're 50, 60, 70. If you are a carpenter, for instance, you are going to be limited by your physical strength.

Are there any types of people who become good CMs?

› People persons and good thinkers.

Describe your role at the CM program at Wentworth.

› I teach and I oversee the advancement of the curriculum and the courses that I have within the program. I have been here for 22 years. I have had four companies before I came here so I've been in the business totally over 40 years.

Is there any other advice that you would give people who want to get into this business, as far as education is concerned?

› Don't stop at the bachelor's degree level. Get graduate education. That's how you can go further in the profession.

How Many Schools Offer Construction Management Education?

A study by the Wentworth Institute of Technology in Boston (www.wentworth.edu) found that construction management programs go by different names. The most popular name, not surprisingly, is *Construction Management,* while "Construction Engineering Technology" and "Construction Management Technology" also account for numerous programs. The most common department or school in which these programs are located is the College of Engineering. This makes it imperative that prospective students widen their search to find the program that is right for them, even though it may go under a name different from construction management.

Be Honest with Yourself

JIM CONNOR

Director, Department of Engineering and Technology

UC Berkeley Extension

If you were counseling someone, say in high school or maybe just beginning college, who has not chosen a major yet, what sort of advice would you give him or her about becoming a construction manager?

› In general, the way Berkeley approaches the topic is we never "promise" a job. We go for the education and the knowledge. It is up to the individual as to how they competitively use the knowledge to succeed. In this case, the construction industry is a business for those people who take a well-rounded point of view, from "Are we building the right thing?" to "Have we gotten a good scope statement and requirements?" This is more than just project management. You are really analyzing the business and the entire environment around it. On one hand, you have the people side: Is it a comfortable, sustainable building? On the other hand, you have government regulations. You also have just plain economics: Should you be optimizing for the construction price, which is classic, or should you be doing other things for its life cycle? So, we try to make sure that people have the proper theoretical foundation.

One year they may be working for an architect, and another year they could be working for a construction company, and maybe a few years down the road they wind up working for a city planning department. Their career will adapt and change. Without the good theory, they are going to have a really hard time. It is the theory that gives you the flexibility over your working career. It gives you the insight to make the best decisions.

We show you how to get a good foundation, and then the certification shows to other people, in a legitimate manner, that you know what you are talking about and you really do know what you are doing. It is a lot more than swinging a hammer.

By the time you have graduated and made the decision to be a CM, what sort of courses have you already taken?

› If you are up on our main campus, you are going to get main campus style courses, including civil engineering and mechanical engineering. Those will allow you to study for and take a California State Professional Engineers Exam. Then, you could say, "I'm a professional engineer, plus I've taken some more of the management style courses. So, not only can I build it, but I can also make good decisions." That is one approach. Then, there is the extension approach. The extension approach tends to be for those people who are already in the industry, who are looking to either change or enhance their jobs. That is where I am. We are the outreach part into the community. We have been quite successful for a number of years; however, we are looking at how to use the CMAA framework as we go through and update our program. We believe that adding in a little bit more on the management side is important. We have a construction management certificate. In it, there are various items such as fundamentals of construction law, managing safety and health in construction projects, and construction project scheduling and control. That's always a tough one.

Are there any other tools that you offer in the form of courses that are really important?

› There is a variety of them. For example, if you were to look in our Arts & Letters side of the catalog, you would see that there are classes on AutoCAD, for example. We have also offered a course on BIM, Building Information Modeling Systems. We offer a wide variety of courses, even the hardcore programming courses where you can learn how to write scripts and so forth, so you can automate your job. We do a few of those also at the lower end, with, for example, Excel, on how to use Visual Basic to automate your spreadsheet. However, we tend not to do the vocational vendor training. If you want to learn how to memorize Microsoft Office 2007, you probably would want to go to an authorized Microsoft type of person. Whereas, if you look at our courses, we will have one on databases because there is a lot of data with buildings: How much did you buy, when, where, why? And with everything else, a database is a good way to organize that information. You probably find that about half the class or so is on the theory of how to make the database and properly design one, and then parts of it are on how to implement using Microsoft, Oracle, or other programs. We always try to make sure students understand how to structure the data so they can keep their ideas well organized.

Any other advice for prospective CMs?

› One thing I would give as advice to people is—and students have a hard time with this occasionally—being honest with themselves. If you are comfortable in using tools or walking on a job site, it is probably a good thing for you to consider looking at CM as a career. However, if it is something that just absolutely frightens you every time something happens, then maybe it is not a good thing. One thing that I normally recommend when I'm working with students, trying to give them advice, is: Do your homework, make sure that you feel comfortable, maybe go out and talk to a few people. It is amazing how many students pick a career just by reading books. Go out and talk to a construction company for a few minutes, and you will get a much more realistic understanding. So, that is one of the key things, if you are going to do it for a while, talk to people in the business to see if that is the type of people you want to be with. I can teach you the theory, but the other part of it is students being honest about what they can do and can't do.

A Typical Course Outline

Courses of study vary considerably from institution to institution. Nevertheless, the following outline of a four-year degree program is fairly typical of a CM curriculum. This is the recommended four-year course of study leading to the bachelor's degree in CM at Wentworth Institute of Technology. Note the real-life work experience sections.

First Year Fall Semester

- Introduction to the Design & Construction Profession
- Construction Graphics
- English I
- College Mathematics I

First Year Spring Semester

- Materials & Methods of Construction I
- Electrical Building Systems
- English II
- College Mathematics II
- College Physics I

Second Year Fall Semester

- Surveying I
- Materials & Methods of Construction II
- Social Science Elective
- Mechanical Building Systems
- Economics

Second Year Spring Semester

- Chemistry
- Structural Design I
- Estimating
- Technical Communications
- Financial Accounting
- Writing Competency Assessment

Second Year Summer Semester

- Pre-Cooperative Work Term (Optional)

Third Year Fall Semester

- Materials Testing & Quality Control
- Structural Design II
- Construction Project Management
- Construction Project Scheduling
- Management of Contemporary Organizations

Third Year Spring Semester

- Cooperative Work Semester I

Third Year Summer Semester

- Advanced Estimating and Bid Analysis
- Structural Design III
- Industrial-Organizational Psychology
- Statistics & Applications

Fourth Year Fall Semester

- Cooperative Work Semester II

Fourth Year Spring Semester

- Construction Project Control
- Construction Business & Finance
- Construction Safety & Risk Management
- Humanities or Social Science Elective
- Power and Leadership

Fourth Year Summer Semester

- Construction Law & Government Regulations
- Labor Relations
- Senior Project in Construction Management
- Humanities or Social Science Elective

End zone expansion project at Dowdy-Ficklen Stadium, East Carolina University, Greenville, NC. CM by T.A. Loving Company. PHOTO: DAVID PHILYAW—T.A. LOVING COMPANY.

The Importance of Accreditation

BRIAN MOORE, PhD
Department Chair and Professor
Georgia Southern University

What advice do you have for high school students thinking about becoming construction managers?

› I would encourage students to go ahead and make sure they get algebra and trigonometry in high school, and maybe even one calculus course. I would encourage them to take AP [advanced placement] courses if at all possible. A lot of student organizations are involved with professional organizations like CMAA, so that's another great opportunity for high school students to get exposure to the industry.

Is accreditation important?

› Most programs are going to try to get accredited because without that confirmation of quality, the marketability of the graduate is dramatically reduced. There is a real consequence for not being accredited. If you are not accredited by ACCE, then you might be accredited by ABET (www.abet.org). In some cases, programs will be accredited by both, but it is the exception and not the rule.

What CM classes do students find most challenging?

› It is probably math and the structures. Every program across the country is a little bit different, but most have courses in structures, and that can be a bit of a challenge for students.

What degree do most students obtain?

› In our school, a bachelor's of science degree in construction management, construction, construction science, or building construction. It depends. The title of the degrees varies across the nation.

Is postgraduate work important?

› If students are graduating and they have not had any experience, they should go out and work for a few years before they consider going for a graduate degree. I think it is important for students to get some experience as they are going through their programs. Do an internship, even if it's not a formal internship. Get to know a company and learn if this is the career path for you.

That said, the longer a person stays out of school, or the longer he is out of school before he comes back, the harder it is to come back. Life gets busy, people get married, they have a mortgage, they have kids to feed, and it can be a much greater challenge to get that postgraduate education. All in all, getting an advanced degree can be very beneficial, especially in today's job market.

What's unique about the program at Georgia Southern University?

› One of the things unique to us is our location. We're 45 minutes from Savannah, a wonderful city, but we are also two hours from Jacksonville, about three hours from Atlanta, and about an hour and a half from Charleston, South Carolina. Yet, we're in a college-town setting. It's quiet, calm, and low crime, so parents can rest easy, but it is also a place where you can get to activity and jobs within a reasonable distance. We also have an international faculty, and the experience they bring to the table helps to broaden the perspective of our students. We have a terrific industry advisory board with active folks that are very committed to the CM program, and we also have a group of associates—alumni who have graduated within the last 10 years. It is a networking opportunity, but also an opportunity to get guest speakers for the program, site visits, and field trips for the students.

The new Exposition Line of the Los Angeles County Metro Rail will be the first light rail line serving Westside L.A. The 15.2-mile line is expected to serve more than 60,000 passengers daily. Safework, Inc., provided safety oversight and other services. PHOTO: GREG METCHIKIAN, SAFEWORK, INC.

Accreditation

When choosing a college or university at which to study construction management, be aware that accreditation of the program is crucial to your success. Not only does accreditation indicate that the courses and faculty measure up to an educational standard, but many companies will not consider your education bona fide unless the program is accredited.

There are many accreditation bodies, including the ACCE, which is a global group that is an advocate of quality construction education programs, and promotes, supports, and accredits high-quality construction education programs (www.acce-hq.org/). According to the group, "Through promotion and continued improvement of postsecondary construction education, ACCE accredits construction education programs in colleges and universities that request its evaluation and meet its standards and criteria. ACCE is recognized by the Council for Higher Education Accreditation (CHEA) as the accrediting agency for four year baccalaureate degree programs in construction, construction science, construction management, and construction technology, and as the accrediting agency for two year associate-degree programs of a like nature."

(The CCMI which administers the CCM program, will accept, as educational qualification, an appropriate degree from any institution accredited by an accrediting body recognized by CHEA. The CHEA website, at www.chea.org, provides a list of these organizations.)

To be considered for accreditation by ACCE, a program in construction education must:

1. Be located in an institution of higher learning that is legally authorized under applicable law to provide a program of education beyond that of the secondary level. Furthermore, in the case of those schools in the United States, the institution must be accredited by the appropriate regional accrediting agency, and in the case of other countries, be accredited by the accrediting agency appropriate for its locale, if such exists.

Have been in operation for sufficient time to permit an objective evaluation by ACCE of its educational program.

2. Offer either a baccalaureate or associate degree program with a major emphasis on professional construction education.

The ACCE has a list of more than 60 accredited bachelor's degree programs and 10 associate's degree programs that have met its standards. These lists are available online at www.acce-hq.org/baccalaureateprograms.htm. Most of these institutions have designated their programs as Construction Management, or they may have similar names, like Construction Technology and Management or Construction Science and Management. Some of these programs, however, have more general titles.

Expansion and modernization of bridges and highways will be a continuing challenge for the United States, and a major focus of professional CM services. This project is a major bridge extension in Glendale, CA. PHOTO: GREG METCHIKIAN, SAFEWORK, INC.

Another accrediting body for you to know about is ABET, formally the Accreditation Board for Engineering and Technology. The group is more focused on engineering programs than on construction management, but many schools that teach CM have the program under their engineering school banner, so ABET accreditation makes sense for them.

Currently, ABET accredits some 2,900 programs at more than 600 colleges and universities nationwide.

Learning the Business Side of Construction Management

J. KEVIN ROTH, AIA

Assistant Professor

Technical Resource Management Program

School of Information Systems and Applied Technologies

College of Applied Sciences and Arts

Southern Illinois University Carbondale

Tell us about your program.

› Our program is just now getting approved on campus. If you come in with a degree—an associate's degree, for example—in construction management, we have a technical resource management program for you. We have a specialization in construction management—which I call professional construction management.

So the prerequisite for joining your program is an associate's degree?

› Yes it is, or at least that number of hours.

How is your program different from others?

› Some of the other programs that you have seen are located in engineering schools, and there is a lot of number crunching involved, but not as much management.

So your program has more of a business emphasis?

› Yes. There is a lot about how to get financing, building permits, bonding—how all this fits together before you put a shovel in the ground.

Why was there a need for this?

› After about 30 years in the business, I see folks who are very strong in certain areas and build these stovepipes in which they want to stay. Well, now everybody has a piece of this, not just construction people. There are investors, bankers, too. I am a firm believer that you cannot manage something if you don't know what it is, and so I'm basically expanding that CM role into what I call a professional CM.

So you take students who already have experience or at least educational experience in CM and make them better business managers?

› Yes. One of the things I know about CM is that it really is a people business. Everyone thinks it is an engineering business or an architectural or a whatever business, but it is really about managing people and managing your clients. This is very difficult in construction.

Merchant Schools

There are a growing number of what are often called Merchant Schools that offer programs in construction management. These are institutions, some of which combine online with in-class instruction, that are often suitable for those who are currently working full time and do not have the time or resources to attend a traditional college or university. Many of these are names that you've heard or seen commercials for, such as ITT Technical Institute, DeVry, and University of Phoenix.

Because of their flexibility, these schools can be ideal for those who may want to get a degree or simply want to take some construction management classes to improve their career choices while they are working at a full-time job.

The Defense Department's Base Realignment and Closing (BRAC) Program involves hundreds of projects worldwide, including the Army Logistics University, Fort Lee, VA. PHOTOS BY ALBERT CRUZ, © MOCA SYSTEMS, INC. ALL RIGHTS RESERVED.

Succeeding as a CM Student

HARRIET MARKIS, PE

Chairperson of Construction/Facilities Management

Pratt Institute

How did you get into teaching?

› I'm a structural engineer by education, and I'm a partner in my own firm. I and my business partner, who also teaches at Pratt, saw a need. Young architects taking the licensing exam needed help. My boss approached me and asked if I would help several of the young people who worked for his best clients, help them with the structures part of the licensing exam.

For architects?

› Yes, for architects. So we started holding these informal mini-classes in the office conference room. It got bigger and bigger and bigger. We decided that we really liked being with young people and explaining what we love best, which is structures. Then, my partner had an opportunity to start teaching structures at Pratt because someone retired unexpectedly. She taught structures to architects. Then, I went over to Pratt too, and I now also teach at Parsons. This story is 18 years in the making. That's how long we have been teaching. We still had our practice, but we would go in one day a week or one evening a week or sometimes two afternoons a week and teach our class. Then, an opportunity opened up here in the construction management department to chair the department. I closed my eyes and jumped, not realizing how much work it is. But, again, I like being with young people who are just starting to discover what they love about the building industry. I find it very invigorating to be with young people who are just starting their professional life.

Tell us a little bit about your CM curriculum. What is special about it that distinguishes it from others around the country and the world?

› The CM program, actually, is a very old program, old by American standards. They offered their first degree in the early 1950s.

Was it in CM?

› It was called *building science*, but it was a CM degree. It was not an engineering degree at that time. It was not an architecture degree. It was a degree in what eventually would be called CM. The culture of Pratt—because there is a Pratt family—was making things. So, the very first school was artisans, builders—people who made things. So, it's no surprise that we have a CM school, an industrial design school, and so forth. Somehow, in the history of Pratt, it became an art school. Artists also make things; there is fashion and fabric and interior design. So, the first students were here on the GI Bill, and even though it's such an old degree, it has never been professionally accredited. The department of architecture is accredited by the NAAB [National Architectural Accrediting Board]. The CM degree has not been professionally accredited. Pratt Institute is accredited, of course, but the program is not accredited. We are in the middle of the accreditation process and if you go to the ACCE site, you'll see we are a candidate for accreditation. It'll be the first-time accreditation for a program that has been around over 50 years.

That will be quite an accomplishment for you.

› Yes, if I live through it (laughs). Yes. It's like nailing Jell-O to the wall sometimes, but I think it is very important to be a part of a professional organization. I think it is very important for CMs to see themselves as professionals. Being professional means keeping up with continuing education because you want to be on top of whatever it is that is going on in your industry. You certainly would never go to a dentist who had not cracked a book since 1954. You would want your dentist to be up to date on what was going on in dentistry. Professionals keep up to date with education; they are responsible for their acts and what they do on a construction site. In addition to that, accreditation forces you to evaluate yourself—what you do well, and what you don't do so well. The self-study is a very big component of that. When you are accredited, you are forced to evaluate yourself on a regular schedule, and I think that is very healthy, for a program to be forced to examine itself and figure out what it could do better and how it should do it better. So, that is what we are doing. In conjunction with this accreditation effort, I went to the ACCE conference, and I was completely blown away by how many really good schools of CM there are in the United States. I couldn't believe it. Every state has one. Many states have several, all sizes. Colorado State University has a student body of 1,000 CMs. Can you even picture that? That is different than a business degree, engineering, and architecture. They are just CMs, and their focus is on heavy construction.

Why do you think CM has become so popular?

› It is not as popular as you might think in New York City, but it seems to be more popular in the middle of the country. I think in New York City, we tend to view it, or parents tend to view it, as a blue-collar profession [for someone] who is paid $200,000 to work in the mud.

Parents are very involved in choices that their high school students make. My parents were not part of my decision, but I was involved in what my children did. It is just the culture we live in. High schools in New York City and the outlying areas don't seem to promote it. Guidance counselors will promote engineering and architecture because they are traditional professions, but they do not promote CM. Maybe it's a deficiency in the way that we are marketing it because, clearly, it is as much of a profession as you want it to be. I had my program advisory committee meeting this morning, and who is on it? CMs from New York City who are helping me figure out what my curriculum should be, so they will hire my graduates. They were all wearing white shirts and ties. So, you could be a field person, but you could certainly be a white-collar person. So, it is not pushed by guidance counselors in the tristate (New York, New Jersey, Connecticut) area as a career path.

Do you think this is a big-city phenomenon?

› I really don't know, it might be. Like I said, I saw Colorado, Nevada, and they all had big programs. I'm completely speculating; I don't know. It's too bad, because it is a really good career choice for the right kind of kid. I won't say it is gender blind, but I do know women in CM. Engineering and CM are good professions for women.

Why is that?

› I think you get a lot of satisfaction. I know that I do when I drive through the city and I see buildings that I know I was part of the team. Not to be sexist, but I think women can multitask easier, and they are focused. I think it is a good profession for women. It is a pity there are not more women who choose that path. I would like to see more women.

What about graduate work; do you do that as well?

› We do not have a graduate CM degree. We do have a master's of science degree in facility management, which is an interesting branch of the business.

That is postconstruction, right?

› Yes. It is the end user, the owner's representative. Again, this is another degree that does not seem to be as well respected in the United States as, say, in Europe, where they have to manage their resources a lot more carefully.

Is there any way you can tell if an incoming student will succeed?

› I think that you have to be someone who—and these would be skills you could learn in four years—is not afraid of public speaking and writing. Who is not afraid to be a team player and sometimes compromise and sometimes stick to your guns. These are traits that anybody who really wants to be a part of the building industry can learn. Some kids are naturals; there is no question. All of these can be mastered with enough practice, though. I think that if you really want it, you will do the things you need to do to get there. Of course, a university can only give you the tools to get your first job; you do your real learning when you are working, but you need to have the right tools and knowledge so that you can be plugged in as soon as you walk into your dream job.

New school facilities, such as this Los Angeles middle school, are being built to tight schedules and budgets, as well as ever-greater attention to sustainability and life-cycle costs. PHOTO COURTESY OF HNTB, INC.

The Phelps School in Washington, DC, had actually been abandoned due to shrinking enrollment. It was renovated and converted to an architecture, construction, and engineering school, with many features—from pre-cast columns with exposed connection plates to exposed and color-coded piping—designed to teach students about engineering and construction. CM by McKissack & McKissack. MAGUIREPHOTO.COM, COURTESY OF MCKISSACK & MCKISSACK.

Scholarships and Mentoring

Many schools offer their own scholarships for construction management students. In addition, many organizations offer scholarships as well. The Construction Management Association of America Foundation, for example, offers financial support to students in four-year college and university programs. Many local and regional chapters of CMAA offer their own scholarships to students in their areas.

To be considered for a scholarship, students must be enrolled as full-time undergraduate or graduate students in a Construction Management or CM-related program for the next school year. Candidates must have completed a minimum of one full academic year and have a minimum of one full year remaining in their course of study. Scholarships are for one year.

Students may compete for the Foundation scholarships in one of two ways:

1. If you have been selected as the top-rated undergraduate or graduate-level scholarship applicant by a CMAA chapter, and have completed a CMAA Foundation scholarship application, your application will automatically be sent to the CMAA Foundation for consideration as a national scholarship recipient.

2. If you live in an area not served by a CMAA chapter, or your local chapter doesn't have a scholarship program, you may complete and submit a CMAA Foundation scholarship application in accordance with the instructions contained in the application package.

Information about the CMAA Foundation scholarship program is available at the Foundation's website, www.cmaanet.org/cmaa-foundation.

Another program aimed at students and young professionals is Construction Manager in Training (CMIT). Designed to introduce and reinforce the basic principles of construction management to entry-level or new employees, CMIT also helps professionals craft career goals that lead them in the right direction towards earning their Certified Construction Manager (CCM) designation.

To apply for the CMIT program, a candidate must meet education and work experience requirements. Candidates may apply to the program while still in college. Applications are reviewed

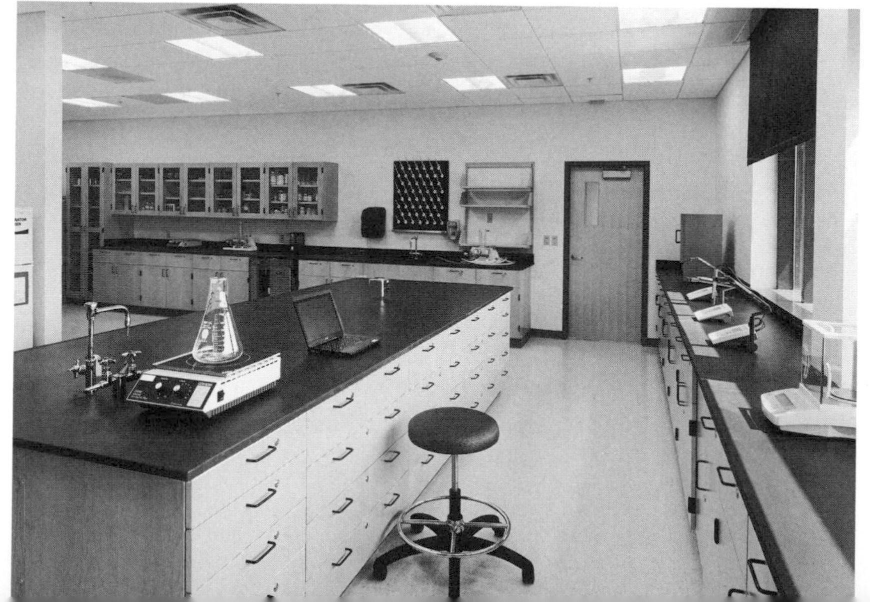

The Science Building at John Tyler Community College, Midlothian, VA, incorporated features ranging from a green roof to diversion and recycling of more than 90 percent of all construction waste. It earned a LEED Silver certification. CM by Gilbane Building Company. PHOTOS BY PHILIP BEAURLINE PHOTOGRAPHY.

by a subcommittee of the CMAA Professional Development Committee. Once accepted, a CMIT candidate must take an online Foundation course that covers the basic concepts of CM work. This course can be taken during the final year of university schooling or shortly after graduation.

The CMIT student must develop a formal five-year strategic plan whose goal is preparation to pursue the CCM credential. Along with the five-year plan, the candidate must submit specific goals for each year of activity in the program. The Professional Development Committee at CMAA monitors each candidate's progress, contacting the candidate at least quarterly to review what tasks have been completed, what goals achieved, and what remains to be done.

A critical element of the CMIT program is mentoring. Each CMIT is paired with an experienced mentor. Mentor applications also are reviewed and approved by a subcommittee of the Professional Development Committee. CMAA has published a complete guide to the mentoring process, designed to help mentors, their protégés, and their organizations understand their roles more fully.

IT Participants Speak about Their Scholarships

SHIRAM JOSHI, CMIT

Cost Estimator & CPM Scheduler in O'Connor Construction Management, Inc., Las Vegas, Nevada

› As a cost estimator, I help architects to develop cost estimates at various levels from Conceptual Design to Construction Documents. As a CPM scheduler, I help general contractors to prepare the baseline schedules and develop them on a monthly basis. My responsibility also includes reviewing the GC's schedules from the owner's side.

I enjoy it because every project I work on is different, from a fire station to large hotel and casino. The nature of my work allows me to work together with architects, general contractors, subcontractors, engineers, and consultants, which I really enjoy.

The biggest value I received from the scholarship, apart from the money, was that, as an international student back then, it was my first opportunity to attend the CMAA National Conference, with a huge number of construction professionals. It helped me know many individuals from my field. In fact, I came to know my employer through that CMAA event.

BRIAN DILLEY, LEED AP BD+C

Assistant Project Manager, Virtual Construction at Layton Construction Co., Inc., Salt Lake City, Utah

› I am currently responsible for the management and creation of Building Information Models for large institutional, healthcare, and government projects. In addition, I am highly involved in the integrated design and construction process on several projects. On these projects, I lead the BIM and Virtual Construction efforts by directing coordination meetings, assisting clients and team members in project visualization, and assisting our field crews to resolve conflicts and forecast construction activities.

The CMAA scholarship provided me with many intangibles that have greatly affected my professional life. To begin with, I was able to network and to meet many people—the majority of them are very influential in the construction industry—whom I otherwise would never have been able to meet. I did this by attending the CMAA National Conference. In addition, the scholarship afforded me the opportunity to achieve the education that I needed in order to open doors to professional opportunities and to improve my family's life.

BRANDON LEBO

Project Manager, Lobar, Inc., a large, family-owned general contractor in Dillsburg, Pennsylvania

› As a project manager, I am responsible for four to five commercial construction projects. My duties include everything from schedule development to subcontractor management. I really enjoy my job because every day brings a new challenge. Whether it is negotiating a change order with an owner or working with the project team to develop ways to beat the project estimate, each day presents a new set of challenges and situations that must addressed.

The biggest value that I got out of the CMAA Scholarship experience was the initial contacts that I made at the annual conference. During the conference, attending the different seminars, dinners, and activities gave me the opportunity to meet many members of the CMAA. For a college student, the exposure that you gain through an experience like the CMAA Scholarship is invaluable. It gives potential employers an opportunity to meet the scholarship recipient on both a business and personal level. Receiving the CMAA Scholarship was one of the most valuable experiences of my college career.

FARRAH FARZANEH

Project Engineer, Webcor Builders, Los Angeles, California

› I was transferred from the Interior/Core & Shell Team on the 54-story L.A. Live JW Marriott & Ritz Carlton Project to join Webcor's Bid Team in their Los Angeles Corporate Office. My duties include creating Building Information Models (BIM) for prospective projects, as well as estimating those projects being bid. I enjoy creating the projects in 3D and 4D prior to the project being built, along with providing take-offs on the project in order to understand the various costs associated. Owners are quite excited to see their vision come to life in a model.

In addition, I work with Webcor's Vice President in Business Development. This starts with the researching of specific project types (Webcor specializes in large projects from $300 million to $2 billion-plus). Once a viable project is identified, I target the project's owner or developer, and work to uncover the information regarding their prequalification process, RFP process, and so on, so that Webcor can qualify to submit their bid for the project. The process requires building a mutual trusting bond with the development team/owner. There are a number of ways to validate to the owner Webcor's commitment to quality, integrity, cost control, excellent safety practices, and professional management of their planning, design, construction, and close-out. Some of those avenues include, but are not limited to, arranging a talk and tour of one of our current projects, so the owner can see and touch the product firsthand; [owners can also] view and discuss the details of one of our past projects, or [I can] meet and greet with them at one of the professionally organized events held by groups like CMAA. I will then coordinate a meeting between the developer/owner and our Webcor executives to consummate the business transaction. Like my father, who was a developer, attorney, and professor, I find these types of business activities extremely exciting and very natural.

The biggest value I received with my scholarship was the mentorship-type attitudes from a large group of leaders/professionals from multiple fields, including construction managers, principles, architects, engineers, contractors, and owners. From the day I walked across the stage to receive this prestigious CMAA award a short time ago . . . to today working in the industry, CMAA members have been as polite and caring as my own family. CMAA certainly attracts those individuals who have mastered their profession and are willing to give back their experiences and advice to students entering the workforce.

AMBIKA ORRILL, PE

Principal Specialist, Integrated Cost & Schedule, General Dynamics, Scottsdale, Arizona

❯ I currently work as an analyst for the Integrated Cost & Schedule Department of General Dynamics-AIS. Through earned-value management, our team analyzes project scope, cost, and schedule in one integrated system. I find the projects that I work on technically interesting (I am currently working on a satellite project) and the performance measurement techniques quite sophisticated.

The scholarship that I received not only provided financial assistance, but also offered me an interface in which I could connect with other professionals in the industry. Whether it was attending the awards ceremony or a CMAA conference, being a scholarship recipient has provided me with many great mentors. Since being awarded the scholarship, I have stayed in contact with CMAA members, and they have all been very helpful in my professional development. I greatly appreciate the generosity of the organization and, having just relocated to Scottsdale a few months ago, I look forward to joining the local chapter and getting more involved with the organization.

Another source of guidance and financial support for aspiring CMs is the ACE Mentor program (ACE stands for Architecture, Construction and Engineering) (www.acementor.org/). Its goal is to "engage, excite and enlighten high school students to pursue careers in the integrated construction industry through mentoring; and to support their continued advancement in the industry through scholarships and grants."

ACE comprises industry professionals such as architects; interior designers; landscape architects; mechanical, structural, electrical, environmental, and civil engineers; and CMs, who attract young people to their professions. These professionals also act as mentors to students.

Since 1995, ACE has presented participating students with $9.6 million in scholarships.

Setting the Standard for Construction Education

MICHAEL HOLLAND, CPC, AIC

Executive Vice President

American Council for Construction Education (ACCE)

What is ACCE?

› ACCE is the accrediting agency for college and university programs that focus on construction management, building science, construction science, and management degrees.

What advice would you give somebody who is looking for a college and wants to become a CM?

› That depends. If they are looking for a degree, my advice would be to look for one that we have accredited. How could I say anything less? We post all that on our website (http://acce-hq.org/), so I would encourage them to look at that list. As far as continuing education, we are not involved too much in that. Again, we are accrediting programs and colleges, not continuing education. We do have a small program where we recognize non-degree programs. We do not accredit them, and we are not a certifying body, but we do recognize them, and there are a few. If they were looking for continuing education, I would ask for their subject, and then I might give them more specific guidance.

Tell us a little bit about how you go about accrediting programs.

› We have a very specific process. We are recognized by the Council for Higher Education Accreditation (CHEA), which has standards that we meet, so that we are the agency for construction-type programs for colleges and universities. We have a very specific program, or process, that includes a program applying and coming in as a candidate working towards accreditation. We look at the elements of the program that would meet our standards and give them suggestions on what they need to do to align and to meet our requirements. We assign a mentor. At some point, they do a complete self-evaluation of their program, which includes the elements that we set up for our standards. Also, we will send an on-site visiting team to confirm that what they said in their self-study is what they say it is. Then, we have a committee process that reviews that information, runs it by the visiting team, and then the board will accredit or not accredit the program. There is a lot more to it than that, and that information is available on the website.

Is there anything else you can suggest to students, as far as your group is concerned?

› Look for accredited programs. Look for specifically accredited programs, not just programs from institutions that are accredited. The difference is that an institution may be accredited but nobody has looked at the specific construction program. The accrediting process is a rigorous process, and it provides value in that standards are established. More and more important elements of standards are established by both industry and academia. We were established by a partnership between industry and academia, and have maintained that balance of influence and decision making, in that our board is made of half practitioners, our accreditation committee has at least 50 percent practitioners, and our standard committee has at least 50 percent practitioners. So, it is not just an academic exercise; it is founded and rooted in the profession. We think that is very important.

If a student were looking at a school, they would want to look at the faculty and see if they have the qualifications to suggest that their people know about construction. I could not imagine graduating from a program that is not accredited. I think students would lose the opportunity to further their profession if they do not graduate from an accredited program.

Continuing Education and Certification

According to the Bureau of Labor Statistics, the majority of employers reported that they offer financial assistance for continuing education; 86 percent offer tuition reimbursement, and 95 percent offer financial support for professional development such as obtaining certification. This is good news for anyone in the field of construction management.

Although certification is not required to work in the construction industry, voluntary certification can be valuable because it provides evidence of competence and experience. Many firms strongly encourage their employees to obtain certification, and may make certification a requirement for promotion to upper-level positions. Again, think in terms of other professions, such as medicine. Would you rather go to an eye doctor who is board certified, or one who is not? Not only do board-certified physicians attract more patients, but many hospitals and group practices require certification as evidence that the physician has continued his or her education and keeps up on the latest advances.

The CMAA awards the CCM designation to practitioners who have the required experience and who pass a technical examination. Applicants for this designation also must complete a self-study course that covers the professional role of a CM, legal issues, the allocation of risk, and other topics related to construction management. To qualify, you must meet stringent requirements for actual project experience over a period of years. This must be "responsible-in-charge" experience, meeting this criterion: Did the decisions that you were empowered to make directly impact the successful completion of the project, and were you directly responsible in charge of construction management services and for protecting the interests of the project/owner?

Applying for the CCM requires a minimum of 48 months of responsible-in-charge experience spread across all phases of a construction project. A qualifications matrix and other eligibility information are available at www.cmaanet.org/cmci.

As you can see, the CCM is oriented to the more seasoned practitioner. It's not something you can get at the beginning of your career.

It's important to note that the major government-issued licenses, like PE, require a certain level of continuing education. Most of CMAA's programs—and those of many other associations—deliver a specified number of continuing education units (CEUs) or learning hours (LHs) for participants, which they can use to satisfy their continuing education requirements. Likewise, renewing the CCM requires recertification points. The goal is to keep continuing education activity going on throughout a career, and there are many ways to do it.

The Los Angeles Dodgers and Chicago White Sox share spring training at the Camelback Ranch complex in Glendale, AZ. The complex includes 10,000 stadium seats, 12 practice fields, and five separate clubhouse buildings, and was executed under a tight schedule with a firm completion date. At-risk CM by Mortenson Construction. PHOTOS BY CARLOS ESPINOSA, 5FOUR PHOTOGRAPHY © 2009.

Did You Play LEGO®?

HENRY KOFFMAN, PE
Director, Construction Engineering
UNIVERSITY OF SOUTHERN CALIFORNIA

What do you ask people in order to gauge their interest in becoming a construction manager?

› One of the things I ask them is about their hobbies and interests. For instance, if they liked playing with Lego's or, in my day, Lincoln Logs™. If they have an interest in building things, having a concrete object rather than an abstract object. Do they get personal satisfaction from that? Do they really like putting things together? Because construction is just a matter of putting the pieces together, making it all work.

Actually for construction management as such, it is really more business skills than engineering skills. So, [I ask] if they like to be able to be part of a team, direct a team. And they have to be goal-oriented to accomplish objectives and be able to work through problems. That's how I would gauge interest.

Discuss the curriculum at your school.

› We're a research university, and we really don't have an undergraduate program as such. We tend to specialize in graduate school. We used to have what we called an emphasis or track in our civil engineering department that engineers could take. I did away with that because I felt civil engineers should have a broad basis of civil engineering in general. We do have undergraduate courses, senior level, that the engineers can take as electives if they want to, so they know that they want to get into the construction area. I did start a minor program about 12 years ago that opened it up to other disciplines, and as a trend in education, it ought to be very interdisciplinary. So we do have a minor program with students from the architecture school, business school, and planning school. I even had a Spanish language major kid come through once. They can come and take a minor program which consists of 23 units in addition to their major. That's really the undergraduate side of it. We really specialize in graduate school. In fact, I think we probably have the largest graduate program now in the world.

Are you talking about just CM or engineering?

› We have the engineering master's of science in civil engineering. Sixteen years ago I started another program called "Master's of Construction Management," which is interdisciplinary, which opened it up to everyone. All you need is an undergraduate degree. It doesn't matter what your major is. Before that, you couldn't get a master's in construction unless you were a civil engineer. In fact, I thought it was such a good idea, I tried to patent it. No one else has done it. Three patent attorneys told me you can't patent an educational program, so I'm losing out on all these royalties (laughs). Every other school has copied me at one point or another. So, in any case, that's my contribution. It's open to everyone. Accountants, even attorneys, come through to get a master's in construction management. It has become very, very popular.

Any last advice?

› Have some aptitude for science and math and also business. Be more analytical thinking along with the social sciences.

JetBlue Airways' new Terminal T5 at JFK International Airport combines state-of-the-art baggage handling and other systems with an airy and modern design. The $735 million, 640,000 square foot project was completed ahead of schedule with CM by Turner Construction Company and Program Management by ProjectConsult, URS Corporation, and Parsons Brinckerhoff. PHOTOS COURTESY OF URS CORPORATION.

Soft Skills

Many colleges are emphasizing the importance of so-called "soft skills" classes for those wanting to become CMs. This is a departure from past years, when schools emphasized the "hard" academic classes such as engineering, math, and drawing. The reason for this change is that the best CMs often are those who know how to get along with people. This is a trait that is *de rigeur* in other business disciplines but, until recently, has been downplayed in construction management in

▲▼▲ Professional CM helped shave about half a million dollars off the budget for the Phoenix College Fine Arts Building, part of the Maricopa Community College District in Phoenix, AZ. The building provides highly varied spaces for visual and performing arts, and was completed without disruption of ongoing college activities. CM by D.L. Withers Construction. PHOTOS COURTESY OF D.L. WITHERS CONSTRUCTION.

favor of knowing how to build. As construction projects become more complex, the need for CMs who not only know how to build but can establish teams, manage people efficiently, and motivate themselves and others is coming to the forefront.

In a paper titled "Improving the Soft Skills of AEC Undergraduates," authors Wilson Barnes, PhD, and Brian Moore, PhD, note:

> As evidenced by the numerous discussions of litigation surrounding construction projects, and more specifically communication and interaction between members of the project team (e.g. designers, contractors, subcontractors, etc.), there appears to be a lack of understanding and, too often, lack of respect between the disciplines. As a byproduct of the resulting inefficiencies, universities in the US are being asked to produce graduates that have 'soft' skills and respectable exposure to the other construction-related disciplines. Interdisciplinary education and improved soft skills are considered to be vital elements of viable construction education. While technical skills have long been considered the most important mission of academic programs, today's graduates having these skills are considered to meet the barest minimum of industry expectations. To meet a higher standard, industry now desires graduates with soft-skill competency and understanding of, perhaps even respect for, other AEC (Architecture, Engineering, and Construction) disciplines.
>
> —PAPER PRESENTED AT THE INTERNATIONAL SYMPOSIUM ON MODERN APPLIED TECHNOLOGY AND MANAGEMENT, BEIJING, CHINA 2005.

Mixing Construction Work and Education

ALI TOURAN, PhD, PE

Associate Professor

Northeastern University, Boston, MA

Tell me about your program at Northeastern.

› The program at Northeastern is not an undergraduate construction management program. It is housed within the civil engineering department, and it offers graduate degrees in construction. It is not unique; there are many programs in the nation that follow this model. The better-known ones are Stanford, Berkeley, and Michigan. All of these big schools do not have undergraduate construction programs. So, people do a degree in civil engineering. Some continue to get a master's in construction, and some go and work for several years and then decide to come back and get a master's or even a PhD in construction management. That is the way our program is set up. It is mainly to cater to people who are either taking the program on a part-time or full-time basis. You'll find out that when you're looking at a rural campus, usually these programs are full-time programs only. On the bigger campuses, in cities like New York, Boston, Los Angeles, because they want to be able to attract working professionals, they open it up to part-time students as well. There are a variety of ways of delivering the material. Courses are offered in the evening. More recently, as in our case, they are offering courses through video streaming. So, students can take the course at their own pace. They do not need to necessarily come to the class, but the tape is available for them. They just review the tape, do the homework, submit their work and projects, and they will be evaluated.

What degree would they receive?

› They will receive a master of science. Some of these schools say master of science in civil engineering, but when you look at the list of courses, they are specific courses to a degree in construction management. Some, for example, say master of science in civil engineering with a concentration in construction management.

What kind of students are you getting? Are you getting people who are already working in construction management?

› There are two types. The part-time students are people who are currently working and taking the program on a part-time basis. These are people who are either working with public agencies, engineering firms like consulting firms, and/or contractors. That is one kind of student. The other type are full-time students. They tend to be a mix of international students or domestic students who continue doing their master's degree after finishing up their bachelor's.

How important is an advanced degree, or how much of a leg up do practitioners get by obtaining one?

› Right now, if you look at the trend in the industry, the idea has been promoted of a five-year engineering education. People call it either minimum requirement master's degree or bachelor's plus 30 credits. It means there is not enough time during the regular four-year program to cover some of the more specialized aspects of engineering. So, if one wants to get some kind of a specialization—let's say construction, architectural, or structural—it means that an undergraduate degree is not sufficient anymore. The people who do this program end up working with various construction companies and/ or construction management firms. You are probably more familiar with those kinds of entities [construction management firms]—companies that have all kinds of engineers working for them. They act as an agency CM for the owner, and sometimes they do claims analysis and project management services, providing it's for a large public agency. That is where the students end up working.

If someone is considering your program, what do you say when he or she asks how your school compares to others?

› Northeastern University is known for its practice-oriented education. I think the main hallmark of the school's fame is its core program. The core program is mainly for undergraduates, but because of that, the courses that we offer have a blend of practice and theory. The students work alternately between classroom and job. For the graduate program, for the past three or four years, we have been offering the option of co-op, and many of our students who come to the graduate program, if we are not supporting them on a research scholarship or something, will want to go out and do co-op work. That actually helps them to find a job later. Sometimes, they continue working with the company they did one or two co-ops with. It is unique to provide co-op at the graduate level.

Where do you come from professionally? What do you teach?

› I did my graduate studies at Stanford in the area of construction management. I have been with Northeastern since 1987. I teach construction management, equipment operations, and we have a research program. Our program is being funded by projects that we get mainly for transportation research. For example, the National Academy puts out a request for proposal that they want to prepare a manual or guideline for the use of design groups. People throughout the U.S. compete for these things. We have been fairly successful at getting them. They come with $200,000 to $300,000 for 1½ or 2 years. We hire students to help us work on these projects, and we submit the final reports. That is a mixture of education and construction.

Have you worked in the field? Or have you always worked in academia?

› I have been a consultant in the industry. I worked as a consultant in the transit industry, so I've worked on several of the larger transit projects. I did analysis of risks. Big transit projects tend to go over budget and have project delays, so with my training, I help model the risks and estimate the problems.

Do you help your students get certified or licensed?

› I teach both at the graduate level and undergraduate level, so most of our efforts are to help our students prepare at the undergraduate level for the fundamental entry exam, the PE exam. We do not have any specific program to help them try to get certified in the CM programs.

3 The Experience of the Construction Manager

THE PREVIOUS TWO CHAPTERS OFFERED AN OVERVIEW of being a CM and the education needed to pursue the profession. This chapter will get into the nitty-gritty of what it's like to be a CM on a day-to-day basis so you will understand what you can expect and what's expected of you.

First, CMs adhere to the *Code of Ethics* laid out in the appendix. In addition, CMAA in 2010 published the latest update of its *Construction Management Standards of Practice*, which was first published in 1986. This document describes industry standards of service and serves as a guide to the range of services that constitute professional construction management. Its goal is to define construction management services without limiting how a professional CM may provide those services.

The *CM Standards of Practice* not only establishes accepted guidelines for CMs but, for those interested in a career, it offers perhaps the best in-depth look at a CM's job across many different disciplines and phases of a project. It describes, in great detail, the role of a CM during every project part.

The essence of good construction management is professionalism and teamwork. The CM, as a member of the team, should assume a position of leadership beginning with the establishment of a management plan. This should not be a position of dominance, but rather of service that integrates the individual elements of the project delivery process into a cohesive program. With this tenet for guidance, the *CM Standards of Practice* manual gives in-depth treatment to ten distinct functions, which are condensed here:

Project Management	Safety Management
Cost Management	Program Management
Time Management	Sustainability
Quality Management	Risk Management
Contract Administration	Building Information Modeling

These functions are not mutually exclusive, but are related and integral components of the construction management process. Each section describes a set of functions and services that are delivered across all phases of a given project or program. These phases are:

- Pre-Design
- Design
- Procurement
- Construction
- Postconstruction

As you read through these functions and phases, you will be struck by the complexity of a CM's job and the fact that the CM is involved in almost every detail of a project.

Remember, it's all about teamwork.

PHOTO COURTESY OF EARL DOC SMITH, EDS BUILDERS, INC.

◀ ▼ Adding a new underground "People Mover" train to a busy international airport is a complex task, as shown in these images from Washington Dulles International. PHOTOS PROVIDED COURTESY OF METROPOLITAN WASHINGTON AIRPORTS AUTHORITY

Coming in from the Field: The Experience of a General Manager

DARRELL FERNANDEZ

General Manager

Parsons Corporation

What does a general manager do?

› For one thing…expenses, anything that is related to overhead. As a general manager, I have to review and approve that. I have to monitor for compliance with federal requirements because we are a federal contractor. I also have to manage overhead budgets and make sure that we are following our budgets and people understand what they can and cannot charge to overhead, because a lot of what we do as a government contractor is highly audited. We have to be compliant. We have staff that does that, but I'm always overseeing it. We look at where we need to be marketing ourselves, involved with different events, and maintaining recognition as an industry leader. As a manager of a company, you have to decide what you can do and what you cannot do.

Are there political considerations as well?

› Yes. Because I'm with Parsons, which is a large company of about 12,000 employees, I'm constantly working in and across different business units and entities. We have to do a lot of communication and coordination with different companies to understand who is doing what and why. Also, we think about what we're doing right and wrong and if we should make changes. I review proposals and approve them before they go out to clients. I help with technical writing of proposals when necessary. In the GM role, you touch a lot of different things. Not that you have total responsibility for any of them, but you touch them all.

There is another side other than the overhead functions. I'll review and discuss or brainstorm schedules. For example, I've been working recently with one of our projects to help them in scheduling. It's a complicated project in a hospital, and there are many concerns about shutdowns and impacts to the operations of an existing hospital in San Diego. I've been down there a couple times meeting with the project and the client. So, I'm helping another business unit because they're short of resources and it helps our overhead.

How long have you been GM at Parsons? Do you prefer this to being out in the field?

› About five years, and I don't like it as much as being in the field. I like it from a family standpoint. I can keep my family very local. My kids are teenagers, and it keeps them in the same school districts. We are not relocating, picking up, and moving, which is typical for a Parsons-type company employee because our projects are national and international. My first few years with Parsons, we did a few moves. My wife said, "We cannot do this anymore. The kids have school now, and we're not doing it." For me, though, the excitement really is being on a project site and seeing things being built, managing budgets and contractors and progress. That is pretty exciting stuff, whereas in the home office doing general management of a company, it is a lot more interfaces with people, politics, and sensitivities. It is not just one motivation to finish a project, and keep a client happy with what we do on site; it's a matter of a lot of different clients and a lot of different potential clients and a lot of internal departments that we have to work with.

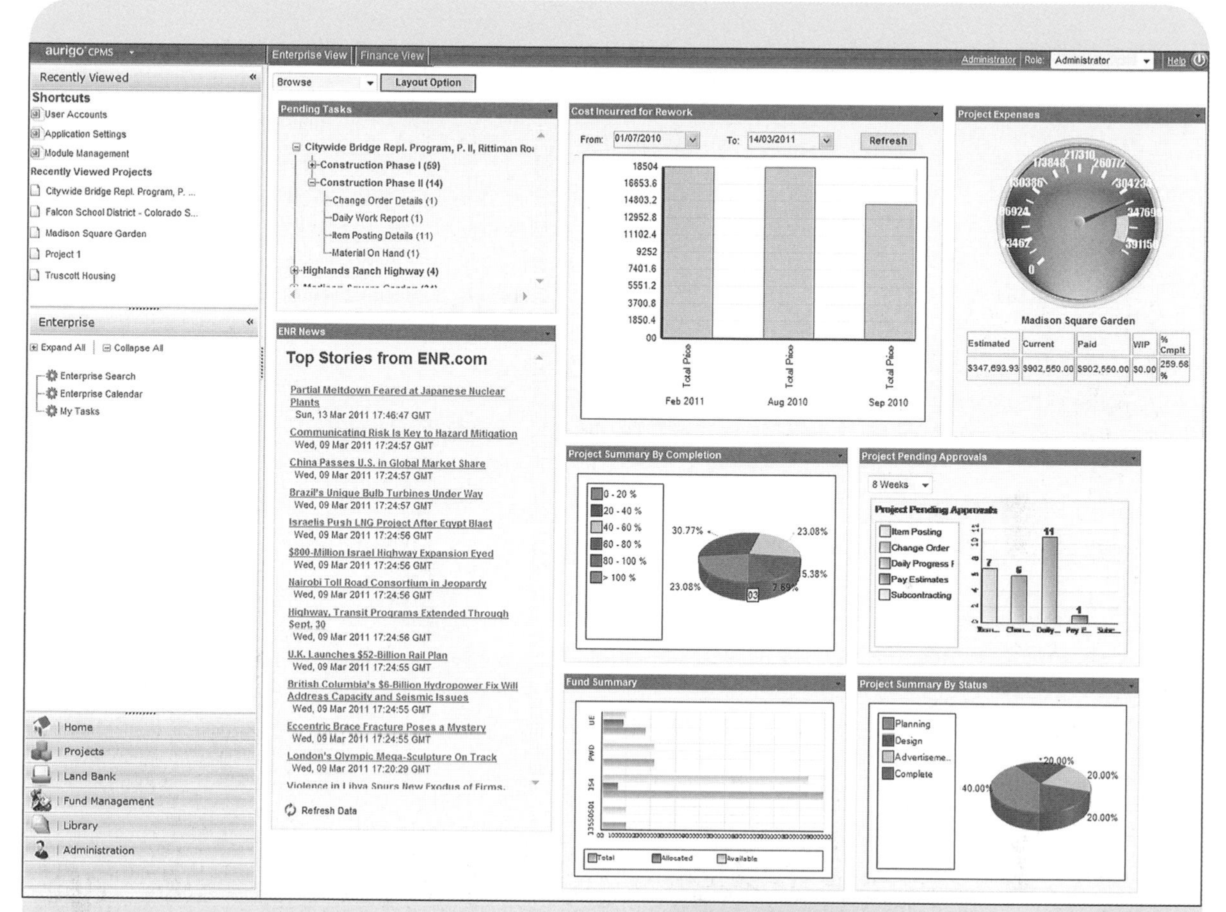

▲ "Dashboards" like this are an increasingly common tool in construction management, as project management software and web-based applications streamline recordkeeping, communication, and project execution. IMAGES COURTESY OF AURIGO SOFTWARE TECHNOLOGIES, INC., COPYRIGHT 2011.

Because, again, Parsons is a big company, we are departmentalized and have to interface with contracts, procurement, project controls, HR, business development people, and senior management, who is always questioning what we're doing and why we're doing it. Sometimes there are challenges with personalities, maintaining flexibility and professionalism, and not being emotional. There is good and bad to that: it is rewarding, but in a different way.

▲ Handheld tablet computer displays put powerful networked resources at the fingertips of field professionals. IMAGE © MOTION COMPUTING.

It sounds like you can really grow by being a general manager.

› Yes, it is not something everybody can do. Some people definitely would not survive.

For a CM, one of the most satisfying parts of the job comes when a project is over and everybody is happy. As a GM, from where do you find your satisfaction?

› It would be seeing a team operate well, seeing people motivated to do a good, quality job, and seeing people enjoy their work. It's leading a team, where you see people happy and laughing, but still working hard at the same time. That is what it is about.

Would you agree that for a GM, soft skills are paramount?

› Yes, they are. One of the hardest parts, to me—I think everybody has a different approach—is the soft skills of dealing with people and recognizing, listening, hearing, validating, and communicating, which is one of the things that makes me effective. Everybody has different viewpoints and opinions. Sometimes that can push people away, or you'll get an A-type personality that can build all types of barriers and nothing gets done, or you can listen, discuss, and draw the best out of somebody. Everybody feels they have taken part in a decision or a process, and that is when people feel they have accomplished something in the home office. You're not seeing

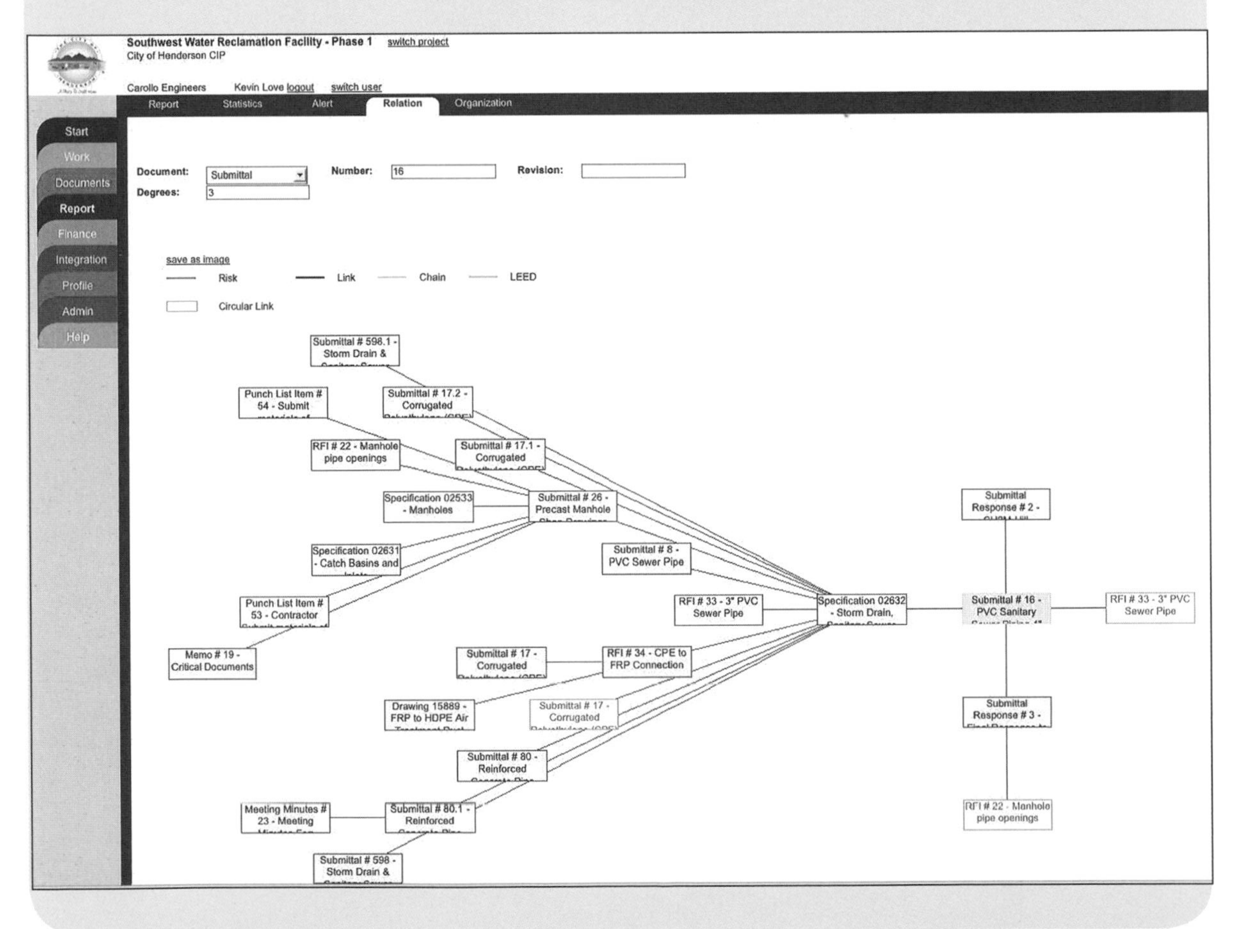

something built, but you're pushing paper, and, in the end, that is part of getting something built. You don't see it physically, which is one of the hardest parts. I'm a construction guy, I've done a lot of construction in college and CM in the field, and then coming into the home office where the only thing you see is paper and email . . . you have to take the best of that and know that it is all part of the process.

Do you think you will go back to the field at some point, or do you think you will do something else after you've earned your GM wings?

❯ I think that when my kids are out of school and we're at that next stage in our life, I think I would be very interested to take on some challenging projects that are in great locations—and I can be very selective. I think I will be back in the field.

Everybody wants to build that shopping mall in Hawaii.

❯ That is exactly right. I visit our projects periodically, and I think. "You guys have it made." We have a lot of work in Hawaii right now, and some of it is right on the beach. One of the projects we have is on the north shore of Hawaii. It's for the Corps of Engineers, doing military training facilities. We went to the officers' quarters to have lunch, and the officers' quarters have glass walls overlooking the ocean and pool where the officers swim laps. I thought, "Wow, I want this job right here."

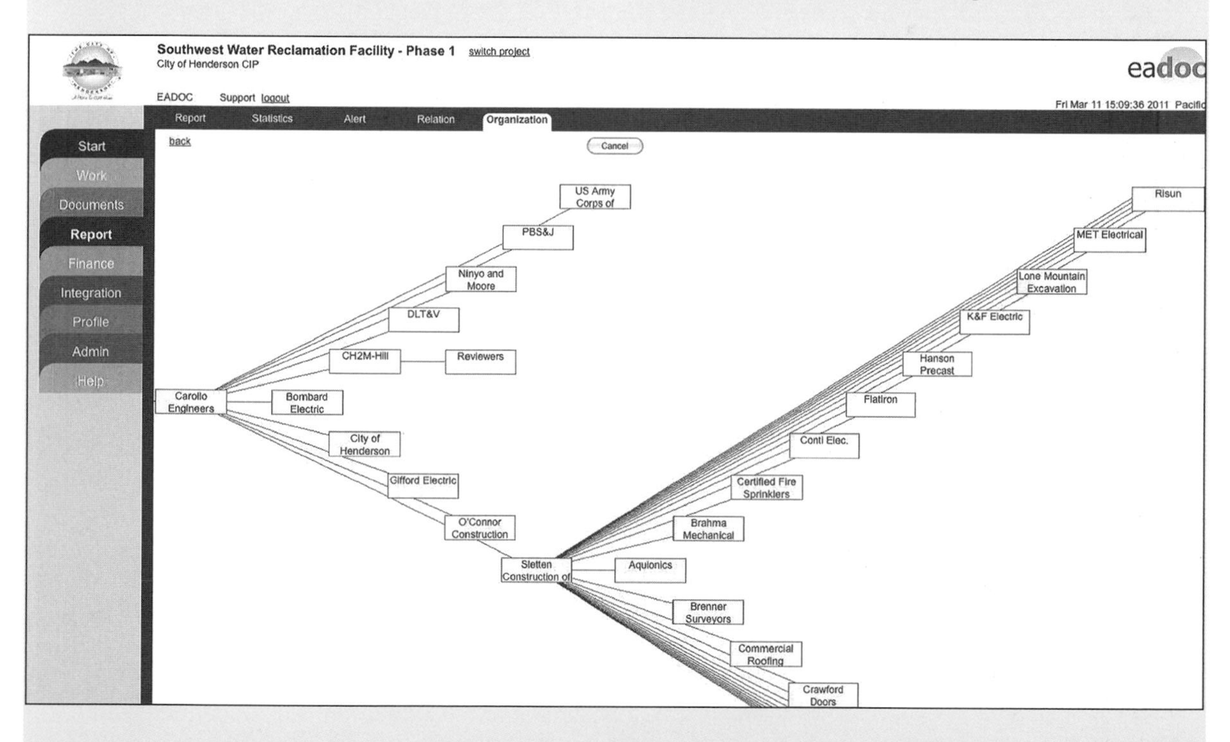

◀ ▲ Large and complex projects can be understood at a glance with tools like project management software. IMAGES PROVIDED BY EADOC LLC WITH PERMISSION FROM CAROLLO ENGINEERS.

Project Management

Project Management is the use of systems and procedures by a team of professionals during project design and construction. In short, project management is the construction job's playbook.

PRE-DESIGN PHASE

During pre-design, the owner must assemble and organize a project team composed of design and construction management professionals as well as other key professional, technical, and administrative staff.

As soon as possible, the project's basic purposes, goals, and criteria of performance—particularly cost, time, and quality—should be determined and documented by the owner and provided to the project team.

The owner should hire the CM and design professionals early in the game. If the CM is hired first, he or she can assist the owner in developing a list of qualified design firms. Additionally, the CM should assist the owner in developing and transmitting the requests for proposals, reviewing the proposals, conducting interviews, evaluating candidates, and making recommendations for the award of the design contract. On the other hand, if the designer is first hired, he or she may assist the owner in selection of a CM.

The CM should work with the owner and the designer to define the project requirements as part of the Project Management Plan (PMP). This document, prepared by the CM, should outline the strategies for building the project. The owner should review and approve the PMP before the project proceeds. This document may then be used to measure the performance of the project team and the overall success of the project. Therefore, it is critical that it be understood by all team members.

The PMP establishes the project's scope, budget, schedule, and environmental conditions, and the basic systems to be used. It defines the methods and procedures to be followed, as well as the basis for claims avoidance on the project. Many conceptual design and estimating versions may be required before a project meets the owner's time, cost, and performance requirements. Once these requirements are established and approved by the owner, however, the team must be committed to completing the project within those requirements.

Typically, the scope of a project is documented by a combination of conceptual drawings, descriptive narratives, performance parameters, and the budget for the project. The type of information and amount of detail may vary considerably, according to the type of project. Documentation of overall cost and time is the CM's responsibility, with input from other team members.

The Project Procedures Manual is best developed as a team effort, assembled and edited by the CM. It should be written so that the responsibilities of the team; levels of authority; communication protocol; and the systems, methods, and procedures to be followed for project execution are clearly defined and understood.

The Manual typically should address:

- Cost controls and the systems required for monitoring and controlling project costs
- The quality control and quality assurance program established by the team, and how it is to be implemented
- The project schedule and how it is to be developed, implemented, and maintained
- Document control and specific project systems, methods, and procedures (i.e., bidding, payments, change orders, submittals, correspondence, reports, performance records, claim resolutions, etc.)
- Functional responsibilities and limits of authority
- Correspondence distribution matrix
- Safety program
- Checklists
- Listing of meetings (i.e., type, frequency)
- Sample forms to be used
- Detailed bidding and construction phase procedures
- Coordination among various prime contractors
- LEED requirements

The CM also usually establishes a management information system that will inform the team about the overall project status and forecast and compare it against the Project Management Plan (PMP). The system provides a basis for managing the project and identifying and evaluating problem areas.

DESIGN PHASE

During the design phase, the team must continually communicate and consult on all issues. As the process proceeds from rough drawings through final design, the team must consider the issues critical to each particular phase, moving from general decisions in the early phase to detailed decisions as the design progresses. The goal is to complete a set of documents defining a project that can be sent out for bids.

The designer has total responsibility for design implementation and execution. Although the designer is responsible for design decisions to meet the project requirements, the owner as well as the CM and other stakeholders can also have decision-making responsibility.

PROCUREMENT PHASE

The goal in this phase is to secure bidders for each bid package. Potential bidders must be qualified, competitive, interested, and capable of doing the work within the time requirements. One of the CM's tasks is to help vet bidders.

CONSTRUCTION PHASE

The CM will work in this phase to expedite and improve the efficiency of the construction process. Prior to construction, the CM should develop a project-specific Construction Management Plan that clearly identifies the roles, responsibilities, and authority of the project team and the procedures to be followed.

POSTCONSTRUCTION PHASE

When the project is finished, the CM's responsibility consists of the following:

- Obtaining LEED certification
- Completion of punch list items not required for substantial completion
- Facilitating owner occupancy
- Assembling record drawings for as-built documentation
- Warranty, guaranty, and operation and maintenance manuals
- Pursuing resolution of warranty items
- Documentation of final pay quantities and costs
- Preparing contract files for transfer to owner
- Final payment and contract acceptance

Maintaining project controls in the field often relies on software and communications technologies. IMAGES © MOTION COMPUTING 2011.

There's Nothing Wrong with Starting at the Bottom

VINAY UCHIL, CCM

Project Manager

PBS&J

Is there a typical day for you?

› Yes, there are typical days, but it depends on what role I'm in. I wear two hats—one is the operations part, and the other is the CM's part. As construction manager, the first thing in the morning I go on-site and make sure everything is going smoothly. For my current project, one of my main roles is to make sure that whatever the contractor is doing, the client is aware of it. It's more about handling the client's expectations, and if there are some issues, putting those issues in front of the client. It's about resolving any issues that come up on a day-to-day basis.

Is your job stressful?

› It's not stressful, it's fun. I enjoy it when I'm in that role, where the client looks at me as a trusted advisor. Another client of mine, a public works client, depends on me for their entire procurement and construction.

What prompted you to become a CM?

› Right from the beginning, I was ready to say what I wanted to do when I grew up in India. This may be very lame, but when I had to make the decision to join engineering, I always wanted to be hands-on, so I was either going to do mechanical or civil engineering. I wasn't looking at any other field. Then, just before engineering, I had a bunch of friends who worked for a contractor. It was a transportation contractor, so they built highways, bridges, etcetera. Every time we would drive by, my friends would say, "This road was managed by this guy," and I would actually know the people. I thought that was really cool, and I wanted to associate my name with a road or bridge that I managed.

It is something you can see. In the beginning there is nothing, and at the end of the job, there is a product that can be seen. That was my first indication that I needed to be in construction. After four years of college, the first job I took was with that same contractor, managing road projects. Then, I came to the United States to do my master's. Even though my master's was in design, I wanted to be in construction. I knew it was going to be tough, because as an international student I was educated in design, and I did not have a lot of background in CM. Luckily, when I interviewed with PBS&J, the man who interviewed me was also from the construction department. I told him I really didn't want to be in design, but in construction, and that I was willing to start at the bottom to get the experience. I got my first job with PBS&J, and I have not looked back.

What kind of projects interest you more than others?

› I like public works. If I had to pick a big assignment from three different clients, I would pick public works because it is the right-sized client for me. If it is a state DOT project, it may be too big a client and you are managing a very small portion of it, so your job can get repetitive. If you're doing a particular type of project, say resurfacing, then it is too small and there is only so much you can do with resurfacing. I feel public works is the right client size for me. It's where you get in and help them plan their next five years financially and also plan all the projects they need to do. You walk them through the entire process of forecasting, design, procurement, and construction.

Most of my experience is in public works, and it has been only in the last few years that I have been dealing with the school district. School construction is fun, but it can be challenging. The guidelines are there to guide you through the process in schools. Many public works planners have no clue what they want, and you have to give them an idea of how to do it. I have been working with rural or suburban planners mostly. Historically, 10 years back, the state would never have built a road or included an intersection, but in the last seven or eight years, counties are trying to do their own work. I find it challenging when I get there, give them a process, and go through the whole thing to actually get their projects completed.

Where do you hope to be in five to ten years?

› That is a very tough question. The one problem with me is I get bored. If the job is not challenging, I quickly want to change the type of work I do. If roads become too redundant for me, I might jump to another client. If I was able to, and I had my option, I would want to manage megaprojects, like billion-dollar projects. The numbers matter because there is a lot more risk and a lot more activity. That is one thing I want to do, but I'm not sure if I will ever get to do it. Also, I want to be a strategist in the field of CM. I don't know what that means exactly; I can't define it very clearly. In my head, it is identifying future markets, developing those markets, and then getting into management. I think management includes planning, developing, and designing. I want to be a strategist, but I don't know how I'm going to get there. Because I want to be a strategist, I have started doing my MBA, so I'll figure it out in the next three years.

Is there any advice you would like to give people who want to be CMs?

› When I started, I started at a very bottom position. I was an inspector for a DOT project. I had a couple of other colleagues who also graduated from Georgia Tech. They often complained, saying, "I can't believe I went to four years of school to do this work [an inspector, for instance], which does not make anything." My suggestion to them was: "You're not here as an inspector, to collect tickets or check the temperature of the asphalt. These tedious tasks are not your primary job function; your job function is to learn the process, learn the construction process, learn how things come together and what needs to be in place for something to be done. Learn what your supervisor does, his job and what makes it better for him." Many people who get into construction in the beginning get frustrated, asking if they went to four years of college to do this, but it's not just that. It's everything around it. Learn how everything combines into a final product.

Cost Management

Effective cost management is simple in theory. It involves the establishment of a realistic project budget within the owner's criteria, but execution can sometimes be challenging as the project progresses.

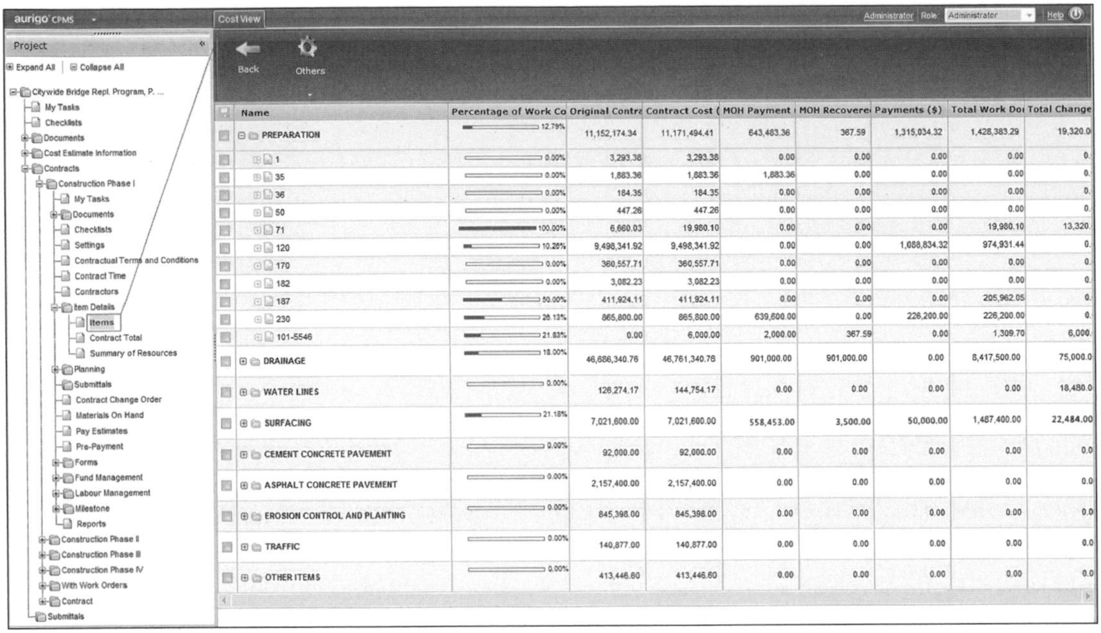

aurigo CPMS | Cost View | Administrator | Role: Administrator | Help

Project

Expand All | Collapse All

- Citywide Bridge Repl. Program, P. ...
  - My Tasks
  - Checklists
  - Documents
  - Cost Estimate Information
  - Contracts
    - Construction Phase I
      - My Tasks
      - Documents
      - Checklists
      - Settings
      - Contractual Terms and Conditions
      - Contract Time
      - Contractors
      - Item Details
        - Items
        - Contract Total
        - Summary of Resources
      - Planning
      - Submittals
      - Contract Change Order
      - Materials On Hand
      - Pay Estimates
      - Pre-Payment
      - Forms
      - Fund Management
      - Labour Management
      - Milestone
      - Reports
    - Construction Phase II
    - Construction Phase III
    - Construction Phase IV
    - With Work Orders
    - Contract
  - Submittals

Back | Others

Name	Percentage of Work Co	Original Contra	Contract Cost (	MOH Payment	MOH Recovere	Payments ($)	Total Work Do	Total Change
PREPARATION	12.79%	11,152,174.34	11,171,494.41	643,483.36	367.59	1,315,034.32	1,428,383.29	19,320.0
1	0.00%	3,293.38	3,293.38	0.00	0.00	0.00	0.00	0.
35	0.00%	1,883.36	1,883.36	1,883.36	0.00	0.00	0.00	0.
36	0.00%	184.35	184.35	0.00	0.00	0.00	0.00	0.
50	0.00%	447.26	447.26	0.00	0.00	0.00	0.00	0.
71	100.00%	6,660.03	19,980.10	0.00	0.00	0.00	19,980.10	13,320.
120	10.26%	9,498,341.92	9,498,341.92	0.00	0.00	1,088,834.32	974,931.44	0.
170	0.00%	360,557.71	360,557.71	0.00	0.00	0.00	0.00	0.
182	0.00%	3,082.23	3,082.23	0.00	0.00	0.00	0.00	0.
187	50.00%	411,924.11	411,924.11	0.00	0.00	0.00	205,962.05	0.
230	26.13%	865,800.00	865,800.00	639,600.00	0.00	226,200.00	226,200.00	0.
101-5546	21.83%	0.00	6,000.00	2,000.00	367.59	0.00	1,309.70	6,000.
DRAINAGE	18.00%	46,686,340.76	46,761,340.76	901,000.00	901,000.00	0.00	8,417,500.00	75,000.0
WATER LINES	0.00%	126,274.17	144,754.17	0.00	0.00	0.00	0.00	18,480.0
SURFACING	21.18%	7,021,600.00	7,021,600.00	558,453.00	3,500.00	50,000.00	1,487,400.00	22,484.00
CEMENT CONCRETE PAVEMENT	0.00%	92,000.00	92,000.00	0.00	0.00	0.00	0.00	0.0
ASPHALT CONCRETE PAVEMENT	0.00%	2,157,400.00	2,157,400.00	0.00	0.00	0.00	0.00	0.0
EROSION CONTROL AND PLANTING	0.00%	845,398.00	845,398.00	0.00	0.00	0.00	0.00	0.0
TRAFFIC	0.00%	140,877.00	140,877.00	0.00	0.00	0.00	0.00	0.0
OTHER ITEMS	0.00%	413,446.60	413,446.60	0.00	0.00	0.00	0.00	0.0

Computer programs enable detailed, up-to-date cost tracking. IMAGES COURTESY OF AURIGO SOFTWARE TECHNOLOGIES, INC., COPYRIGHT 2011.

PRE-DESIGN PHASE

Prior to developing any construction cost data, the CM must become familiar with the construction site and the environmental factors—such as upcoming weather conditions—that will affect costs. In addition, the CM must assess the local economic conditions (Are resources such as skilled labor available? How about materials and heavy equipment?) and investigate other potential risks.

A CM should study past construction projects in the area to get an idea of costs so he or she can model the upcoming project. Using these data, a cost forecast can begin to be built. Based on the owner's goals in terms of performance, quality, and time, the CM develops an estimate of the cost of construction. This information is incorporated into the PMP.

Because the CM is dealing with rough estimates, it is not uncommon to build in an additional 15 to 25 percent. It's important for the CM's credibility that budget estimates include the assumptions upon which the estimate is based, so the owner can appreciate the way in which the cost was determined.

DESIGN PHASE

The approach to managing costs during the design process should be proactive rather than reactive. Active participation of team members and coordination of the CM with the design team in providing advice can significantly reduce the need for redesign due to cost overruns. Once the construction budget is approved, the CM provides ongoing cost management services to make sure that budget is being followed. A uniform cost estimating framework is established and maintained from inception through the pre-design, design, bid and award, and construction phases of the project.

When developing estimates of construction cost during the design phase, the CM needs to carefully study all the specifications, because they can provide information that may not be shown on the drawings but that may have significant cost implications.

Saving Money through Technology

CONSTRUCTION

As part of the overall financial control during the construction process, the CM establishes and implements a *change order control* system, which keeps track of payments for unexpected changes in work. A change order is a written agreement or directive between contracted parties that represents an addition, deletion, or revision to the contract documents; identifies the change in price and time; and describes the nature (scope) of the work involved. Change orders can occur because of construction errors, owner's changes, or unpredictable conditions such as weather or natural disasters.

Part of the system is a method to decide how much something should cost, considering that it was not part of the original plan. The only fair way to do this is by obtaining and studying supporting data of true costs. This is a critical job of the CM to keep cost overruns from getting out of control.

CONSTRUCTION PHASE

During the construction phase, the CM should monitor cost management procedures until the job is finished. It's up to the CM to make sure that all underpayments are assessed to the proper party and that any overpayments are doled out fairly.

An Appreciation for Public Service

WALTER FEDROWITZ, CMIT

Project Engineer

FLUOR

What's your job at FLUOR?

› I'm technically a project engineer. I work for FLUOR currently as the quality control manager on a design-build project in Richmond, Virginia. It is a road and bridge job; it's $38.5 million for four bridges on a mile and a half road. My job is to make sure that we are meeting all of the standards, specifications, contract documents, and so on. On the job, I'm the quality control manager. If I were to be on another job, I'd probably be a project engineer.

What got you interested in being a CM?

› I graduated from Virginia Tech in December 2007. When I went to school, my first year of engineering was a general track, and at the end of that first year and the beginning of my second year, you choose your major. I did an internship in mechanical and electrical engineering. It was good, but at the end of that summer I had the opportunity to tour the Woodrow Wilson Bridge [which spans the Potomac River on the Washington Beltway]. After an hour-and-a-half tour, I was sold on switching to civil engineering with a focus on CM. I interned on that job the following summer and loved it. I interned with a bunch of other companies the summers after that. Now that I'm full-time, I know I made the right call. I love the teamwork, everybody working together in a different form or fashion to build a job for the general public.

Is there an aspect of public service that you appreciate?

› Yes. For example, a software engineer will design a program that people get to use. I'm building roads and bridges that people get to travel on. Hopefully, it is safer than the previous bridge. They are getting from point A to point B quicker and safer. At the end of the day, you can actually see the product that I built. Not all engineers get to say that. It's a large structure that everybody can see. The job I'm on is the Richmond Airport Connector Project, connecting Route 895 to the Richmond Airport.

Since you are relatively new to the industry, where do you see yourself in five or ten years? Do you have a plan for yourself?

› I do, and it continually changes. It's being on a heavy civil project, maybe not as a project manager, depending upon the size, but in some form or fashion managing the project. Right now, I work for a contractor, so I manage construction day-to-day and hands-on. I do the means and methods to figure out ways to get jobs built quicker, faster, and safer. I don't know exactly what position it would be, but it would be as an integral member of a team on a large, heavy civil project.

On a typical day, I come in and review test reports from the previous day. I go out in the field and check on a few areas of work I'm responsible for. When I come back in, we have a meeting and set the plan for the next day. I make sure all the foremen have what they need, review safety and quality production for the next day, and do more reviewing and creating reports. In the afternoon, I'll go out again to check on the field. I'll make any field decisions that need to be done. Photos, RFIs, or anything that changes in the field, I'll get the information to pass it on to the designer. The end of the day is just closing the loop, working in a more office-type atmosphere with documentation. Every day is different. This is my typical

Sound construction management has significant impacts on important community resources, such as the Columbia Heights, MN, Public Safety Center. CM BY EDS BUILDERS, INC. PHOTO BY EARL DOC SMITH.

▲▶ Heery International was CM for this major community resource, the Georgia Aquarium in Atlanta. PHOTOS: COURTESY HEERY INTERNATIONAL.

day, but some days I'm in the office 80 percent of the time, and some days I'm in the field 80 percent of the time. That is what is exciting about this field; every day is different.

What are some of the tools that you particularly find useful in your day-to-day work?

› Sometimes, it is as simple as a set of plans and a calculator. We use a software program called CMIC; it is an all-inclusive project management, financial management, equipment management software tool. We use that to keep track of everything. One of the basic tools is just a standard specifications book. You use that to make sure you are supplying the owner with the product they are expecting.

Time Management

Construction management involves the management of three basic project parameters: cost, time, and scope. All CMs recognize that these three parameters are closely linked and that a change in one can affect the others. This relationship is sometimes known as the *triple constraint theory* and is often depicted by a triangle. An increase or decrease of one side affects the lengths of the other sides.

CMs must manage these three parameters and maintain the proper balance. Time management is an integral part of the CM's responsibilities on a project because, as the adage reminds us, "Time is money."

Generally, the CM's responsibilities related to time management are:

Ensure that the project team develops a project plan that considers time.

Ensure that the project team develops a schedule to both plan and monitor time on the project.

Guide the project team as to the appropriate form and content of the project schedule.

Act as the leader of the time management and scheduling effort.

PRE-DESIGN PHASE

Along with the owner, the CM will develop the *master schedule.* This time roadmap will include all of the sequences and orders needed for the project. Once the owner accepts the master schedule, the CM produces the *milestone schedule*, which highlights key events from the master schedule. This is how the CM will determine if the project is on time, running late, or ahead of schedule.

DESIGN PHASE

The CM's responsibilities related to time management during the design phase include: maintaining the master schedule, working with the designer on a schedule for the design phase, monitoring the design phase, developing the pre-bid construction schedule, and preparing and distributing reports to the owner and others.

PROCUREMENT PHASE

At the pre-bid conference, the CM, designer, and owner will work together to explain to potential bidders about the project's timeline. Ultimately, any successful bidders will become part of the scheduling process, and the CM will show them the milestone schedule as well. From this, the successful bidder will prepare their construction schedule. All of this will be spelled out in the contract.

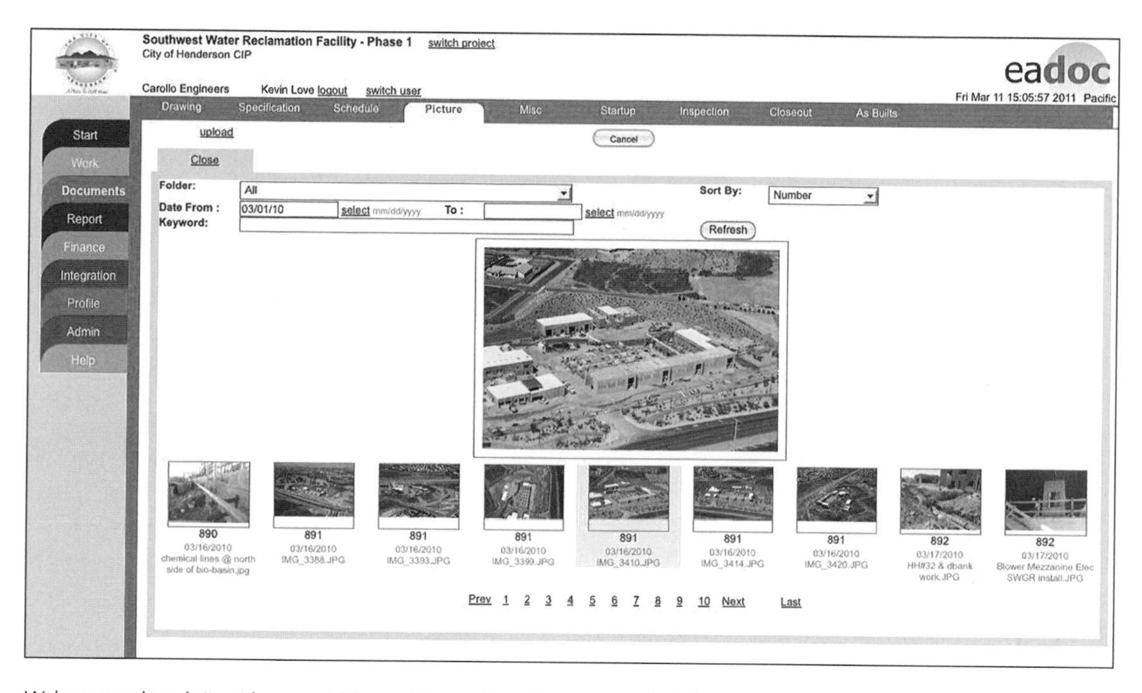

Webcams and worksite video are widely used to monitor adherence to schedules and verify site conditions at specific times during a project's life. IMAGE PROVIDED BY EADOC LLC WITH PERMISSION FROM CAROLLO ENGINEERS.

CONSTRUCTION PHASE

During the construction phase, the CM is responsible to ensure that a schedule is in place for each phase of the project and for each stage of each phase. The CM also will typically be tasked with ensuring that the contractor fulfills its obligations under the contract as it relates to schedules and with recommending appropriate actions to the owner in the event the contractor fails to meet its obligations.

The CM also plays a central role in the development, acceptance, implementation, and monitoring of the *baseline schedule*, which allows the CM and others to measure their progress.

POSTCONSTRUCTION

The CM may help fill a building quickly by developing an occupancy plan that offers the owner a smooth and orderly transition into the completed project. Once the occupancy plan is accepted by the owner, the CM should incorporate it into the master and milestone schedules for the project.

The Sole Proprietor

TODD NIEMANN, CCM

President/CEO

TWNIEMANN, INC.

What is a typical day for you as a sole proprietor construction manager?

› I'm up very early and usually on the computer taking care of emails. I'm very active in CMAA as well; I'm on a couple national committees and also local chapter committees. So, there are a lot of emails, and I handle those in the mornings. I do all my own invoicing, business paperwork, and daily reports for the projects. I keep notes during the day, and in the morning I'll enter it into an Excel worksheet for the daily logs. Basically, I handle all the business in the morning and then I head out to the project.

Today, for example, I'm going out on a storm drain contract. My preference is really park sites; they are my specialty, my expertise. Everybody appreciates a park; no one really cares about a storm drain. On the job, I make sure the contractor is doing the work according to the contract documents and the project specifications. I deal with the residents or any other utilities or agencies that may be involved. I'm out on the site and I represent the city. Right now I'm working for the city of San Juan Capistrano [California]. I'm their eyes and ears out on the project and represent them. If there are any resident complaints, I deal with the people and make sure they are happy. I make sure the work is going according to plan and specifications. I keep tabs on everything and make certain to keep good notes of all the people out there, the equipment, and the progress that has been completed.

Did you work for a company before you went out on your own?

› Yes, I worked for a large civil engineering firm in California. I was vice president and ran their CM division.

Why did you go out on your own?

› When I worked there, I had a lot of independent contractors (ICs) who worked for me. Because construction projects come and go, I would hire these IC guys for a three-month project. I saw that these guys basically just had a car, cell phone, laptop, printer, and digital camera; they were on their own and did business. It was a simple deal, and it always intrigued me. I've always thought of doing it, but I would never have actually jumped ship while I was there. I was a corporate guy, a private stockholder; I thought I was going to retire there. Honestly, I thought I wanted the corporate job and I was moving up the corporate ladder. I actually started running one of the offices as an office manager. I started dealing with all the HR stuff that goes on with running an office. I wasn't *on* the project, although I was responsible for the guys and made sure they had work. They were on the project, and they were all billable. I had to stay billable myself, but I wasn't on the project. I didn't really like that; I didn't feel good about having to bill projects when I wasn't really on them. Also, I've grown up in construction. My dad was a contractor, and I've been around it my whole life. I started as a laborer in construction, and the part that I enjoy most is being on the job and making things happen. Every job is different; there are always issues, always something that comes up that wasn't planned on, or a changed condition. There is always something you will encounter. I really like the nuts and bolts of trying to figure things out, coming up with better logistics and different ways of doing things, and the hands-on aspect. I was getting away from doing that kind of stuff and doing more of the corporate management.

I was contacted by a skate park designer who wanted to start a skate park construction company. He didn't want to just design them; he wanted to build them. The design didn't pay very much, but he saw that the contractor doing all his jobs was getting rich, and he wanted to capture some of that. I was born and raised in California. I am a sixth-generation native, and I've grown up on a skateboard. I still skate; my son and I both still go to the skate park. It really intrigued me. It was a great opportunity, but a tough one. That is what made me jump ship; I would not have done it unless something like that would have come along.

What is the biggest challenge of being on your own?

› It's getting the work. A lot of times clients will hire a large CM firm because they have resources. If one guy is out sick, they have another guy they can throw in his position for the day. Sometimes a client does not want a one-man shop, because they want more resources.

Quality Management

Quality is crucial to any successful project, and it's the CM's job to make sure that quality management is an inherent element of his basic service, even if it is not specifically addressed in his contract. The CM should encourage the owner to develop and implement a comprehensive Quality Management Plan (QMP) as one of the first project tasks to be undertaken.

PRE-DESIGN PHASE

The goal during this phase of the work is to establish a program of quality management that will last throughout the life of the project. The CM should meet with the owner to clarify the expectations, goals, and objectives of the quality management program. He also should explain the costs and benefits.

DESIGN PHASE

During this phase, the CM should ensure the implementation of the QMP. One of the most important aspects is a Quality Assurance Plan (QAP)—part of the QMP—which should be followed by the designer, including the systematic reviews, which demonstrate that quality control (QC) activities have, in fact, been undertaken. The CM provides oversight review of the design professional's QA efforts on behalf of the owner.

The CM also should develop quality management specifications in which the contractor's QA/QC responsibilities are identified. On larger projects, it is also desirable to require the contractor to implement a written QMP. The CM should confirm that these specifications are included in the contracts.

CMs planning new projects at Fort Belvoir, VA, used Google Earth 4D models to assess the best sequence of construction, as well as impacts on surrounding structures and activities. SCREEN SHOTS COURTESY OF BELVOIR NEW VISION PLANNERS GOOGLE EARTH 4D MODEL.

PROCUREMENT PHASE

The CM establishes the goals for the procurement phase as a part of the QMP. The master schedule, as outlined in the "Time Management" section of the *CM Standards of Practice*, also should be consulted. The CM should review the master schedule procurement cycle for advertisement, bid, and award, together with any special approvals during the award cycle, to ensure that the schedule reflects market conditions. All other tasks such as procurement planning, advertising and solicitation of bids, selecting bidders, pre-bid conferences, bid opening, pre-award conference, and finally, the contract award should all be tested against the QMP.

CONSTRUCTION PHASE

As in the procurement phase, elements of the construction phase should be accomplished while adhering to the QMP. These elements include: preconstruction conference, construction planning and scheduling, inspection and testing, reports and recordkeeping, work changes, document control and distribution, handling nonconforming and deficient work, progress payments, final inspection, and final acceptance.

POSTCONSTRUCTION PHASE

After the project is built and all CM services are nearly complete, the CM may review and discuss the overall quality management of the project with the owner. CMs usually hold discussions with the owner, which allows both sides to learn from their experience with the aim of improving quality in future jobs.

BIM rendering of mechanical dewatering facility including 2-meter belt Filter presses and associated equipment, Crum Creek Water Treatment Facility, Springfield Township, PA. IMAGE COURTESY OF HATCH MOTT MACDONALD, 2011.

Accelerating Innovation in the Field

KEVIN DURHAM

VDC Technical Specialist: Virtual Design & Construction

AUTODESK

Describe the importance of technology to CMs.

› Specifically to BIM (Building Information Modeling): Being able to virtually simulate, understand, and clash-detect a building before you ever put anything in the dirt is the real importance of it. Being able to walk around the complete building, run clash detection on it, schedule against things to make sure you have the right lay-down areas, equipment, and materials on site. We're seeing a lot of work around safety planning with our tools now to make sure scaffolding is in place, bridges are in place, and other equipment is in the right areas. We're seeing it across the whole scope of the project where people are finding uses of BIM-like tools to enhance their projects.

Some states, such as California, are requiring that every single item be done with BIM, even things as simple as doorknobs not touching other doorknobs. What you are seeing?

› California, obviously, is out there in front of everybody else. I travel the country and Canada, and depending on where you go there are different levels of acceptance of BIM and what is being used in the industry. Some areas are still in the infancy of discovering what BIM is. Other areas are definitely getting there. Everyone is exploring and understanding. I don't even know that Autodesk understood how much could be done. It is amazing how much innovation is being done out in the field now with our tools and other tools, to facilitate some of these things. I don't know if that was on everybody's roadmap when we first started going down this path.

What sorts of things are you seeing that you didn't think would happen or didn't anticipate, and that were a surprise to you?

› Actually, the logistics and safety planning is something I have been digging into for a while. To be able to create safer job sites from a safety point of view and from a logistics point of view, using it to understand where and how you can store your materials better on a job site, so you do not have to take up as much space next door or in a warehouse.

The other big one is the reinvention of prefabrication in the construction industry. I think everybody has been scared of prefabrication because it was some guy who was building a house and he could prefabricate four walls and deliver them on-site, and that is all he did. Where I am seeing innovation is in using BIM models to understand how they can break up prefab materials into "bite-sized chunks" that will fit on the back of a flatbed or other vehicle. Miami Valley Hospital in Dayton, Ohio, prefabricated racks for the hallways, and also the restroom modules. They were able to reduce their labor rates by 20 percent by building them on the shop floor rather than in the field. Their plumbing quota went from 200 feet per day to 600 feet of plumbing per day on a job site because they were able to prefab in the warehouse. They build racks, get them ready, then fly them up into the building itself and bolt them into place. They use quick connects between these racks, so all the electrical components literally just hook up together and you're done. They did the same thing with the bathroom modules. They actually did little recesses in the concrete floor, so the bathroom modules slide in on dollies and drop right into the recesses on the concrete floors; hook up the components, and you're done. That, to me, is something I hope we see more of—the prefabrication side of it.

Looking out five to ten years, where do you see things going? What other things are people asking?

› I think the big push over the next five to ten years, and I hope it is earlier than that, will be around the operations of maintenance and facility management with these tools. We are feeding boatloads of information into these BIM models at an early stage, and it would only make sense that the owner gets the BIM model and can own and operate their building or facilitate their building from those models, rather than from a bunch of rolled-up drawings in the back of somebody's office.

In commercial spaces, like shopping malls and office towers, there is always some level of facilities management going on. What I think has happened is that there has been no innovation around it; the tools are pretty archaic. I just don't think it is on the forefront of anybody's mind because there hasn't been a lot of movement around it. But I think it's the next surge.

Contract Administration

One of the most important aspects of being a CM is the administration of contracts. Contracts detail all parties' tasks and responsibilities. If there is a dispute during a job (and these are common), the contract usually serves as the arbiter.

PRE-DESIGN PHASE

The CM should develop procedures for recording and controlling the flow of submittals by the designer for approval by the owner. The CM should establish the systems and procedures for communications among the owner, designer(s), and CM over the course of the project.

DESIGN PHASE

The CM's job here largely consists of implementing a system for information to flow to all project team members concerning the development of contracts. During the entire design phase, the CM maintains a process of review and consultation among team members. The CM will keep notes of meetings and make sure they are distributed.

Once the master schedule and milestone schedule have been prepared, the CM initiates a schedule maintenance report. This report is intended to monitor the project schedules and compare the actual progress, particularly of critical dates, against the scheduled progress.

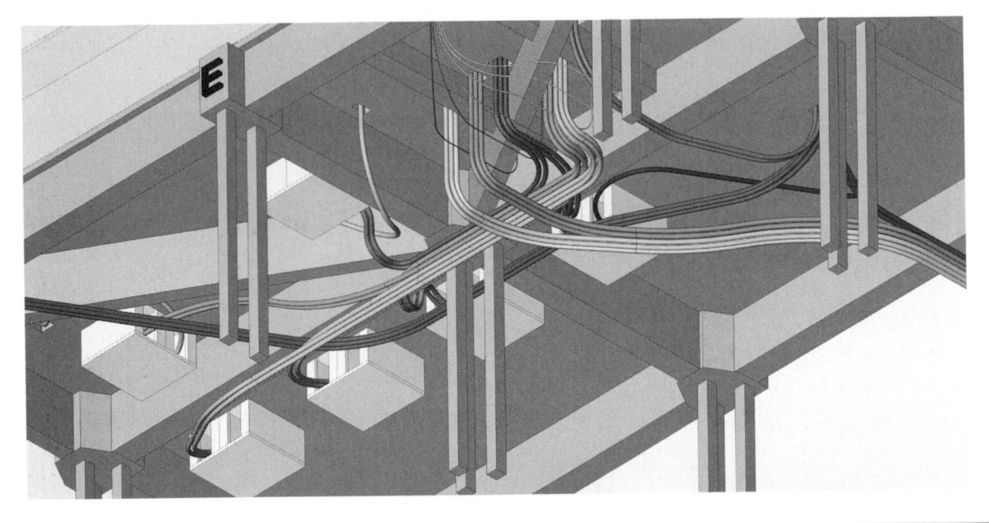

"Clash detection"—finding places where different building components bump into each other, or where existing conditions make it impossible to execute a design—is one of the key benefits of BIM. Clashes like these are much easier (and less expensive) to detect on a computer screen than in the field!
TOP IMAGE COURTESY OF VANIR VIRTUAL PROJECT MANAGEMENT GROUP. OTHER IMAGES COURTESY OF M.ARCH ARCHITECTS.

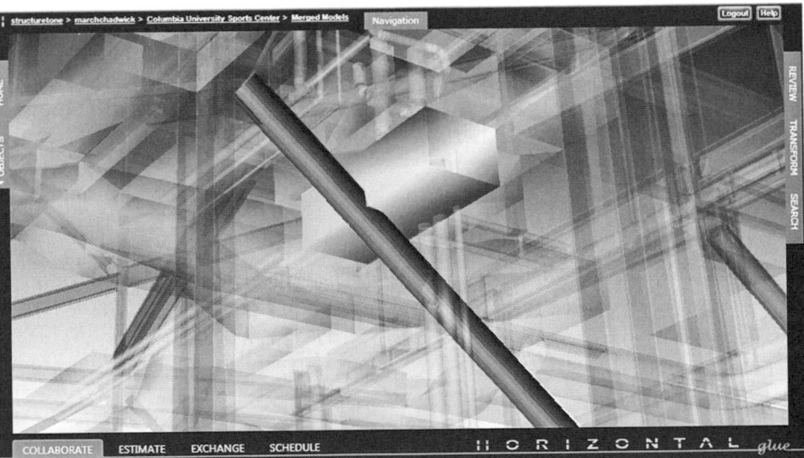

The CM also prepares a project cost report to compare the budget for the project to the actual costs incurred and the forecast to complete. Initial project costs are conceptual, but become more accurate as the project is defined and then constructed.

PROCUREMENT PHASE

The CM's job is to assist in securing for each bid package a sufficient number of bidders, including subcontractors, who are qualified, competitive, interested in the work, and capable of doing the work within the time and budget. *During the bidding process, the CM's role is robust.* The CM may develop a contract breakdown for each contract on the project, which will consider availability of design information, schedule, and local contracting practices.

The CM should assist the owner in developing the list of potential bidders and in prequalifying bidders. Project bidding may be open to all interested bidders or only to prequalified and approved bidders, depending upon the owner's wishes. In an open bid environment, the CM should evaluate the bids for competitiveness, responsiveness, and ability of the bidder to do the work. The CM should also confirm that the bidders are financially strong. For closed bidding, the CM should develop the criteria for bidder selection in consultation with the owner and the designer.

The CM may be called upon to conduct a telephone and/or a written campaign to generate interest among bidders. If the owner wants to have open bidding instead of a selected bidders list, the CM should assist the owner in the preparation and placement of notices and advertisements in trade journals and newspapers.

The CM also may be responsible for conducting pre-bid conferences—forums to explain the project schedule, access to the site, time constraints, owner's administrative requirements, and technical information.

During the actual bidding process, the CM may assist the owner in recording bids and making sure that they met the deadline. The CM may assist in evaluating the bids for completeness, responsiveness, and pricing. The CM may also conduct post-bid interviews to discuss the proposed contract with the winning bidder. All parties want to make sure that there is a clear understanding of project scope, and to discuss any bid alternatives the bidder may have submitted. The CM should confirm the absence of any errors and inform the bidder of the permit requirements, as well as the required insurance documents and labor affidavits, quality issues, and any special requirements.

After the owner approves the successful bidders, the CM may be called upon by the owner to help execute construction contracts.

CONSTRUCTION PHASE

During the construction phase, the CM monitors the progress of contractors in securing and maintaining proof of insurance, building permits, insurance, labor affidavits, and bonds. The CM should call a meeting of the project team and others to discuss the requirements of the contract and the contractor's approach, and to review the administrative and other reporting procedures required prior to a Notice to Proceed. The CM should prepare an agenda and conduct this meeting prior to the contractor(s) moving onto the project site. The CM should issue the Notice to Proceed to the contractor once it is confirmed that there are no outstanding issues that could delay the start of work.

As previously discussed, one of the main jobs of a CM is communication with others. This is pronounced during this phase, which includes setting on-site communications procedures, conducting site meetings, distributing documents, and reporting on-site activities such as the number of workers and visitors, significant materials received, project delays, and a general description of each contractor's activities.

The CM should also continue issuing reports begun during the procurement phase. These reports are:

- Schedule maintenance reports
- Project cost summary reports
- Cash flow projection reports
- Construction schedule reports
- Progress payment reports

POSTCONSTRUCTION PHASE

During the course of the project, the contractor is required to submit maintenance manuals and procedures for operating equipment and systems. Prior to final completion, the CM should coordinate the compilation, organization, and indexing of these materials and bind them into document sets.

When the project ends, and contractors have completed all their work, the CM initiates *contract closeout*, which consists of:

- Certificate of substantial completion
- Completion of punch list work
- Final lien waivers
- Guarantees/warranties
- Final payment application

All significant reports that have been issued during the design and construction phases should be summarized in a final project history report.

Safety Management

What is the CM's responsibility for project safety?

This is not an easy question to answer because safety management should be a collaborative effort, and, as such, *some* responsibility for safety is widely shared. All project participants, for example, have a duty to call attention to unsafe conditions as a key step in preventing injuries to themselves or others.

In addition to this basic ethical requirement, other responsibilities for job site safety derive from laws, regulations, and contracts.

The CM should counsel the owner in creating contracts and processes so that:

- Responsibility for specific risks is assigned to the party most able to control and mitigate those risks.
- A "safety culture" on the project makes it clear that all participants are expected to report known hazards to the appropriate individual or entity responsible for the involved work, as well as to perform their own activities in compliance with laws and regulations.
- Appropriate liability protection is provided to those parties responsible for monitoring construction activities on behalf of the owner.

Among other functions, the CM should be alert to inconsistencies in contracts. Laws and regulations take precedence over contract terms. Generally, the CM's obligation to provide services related to safety varies substantially from project to project and must be clearly specified in the contract between the owner and the CM. The CM should also review the contractor's agreement with the owner to ensure that there is no language specifying CM safety obligations or responsibility that is not clearly defined in the CM agreement.

In the case where an owner has a well-established safety program/organization, the owner can provide a safety coordinator to perform the noted project safety functions, and to work with the CM. If neither the owner nor the CM has an established safety program/organization, an outside safety consultant may be brought in.

PRE-DESIGN PHASE

The CM should review the contractor's contract prior to procurement to ensure that there is adequate language for managing safety. *Project safety should be considered a process that is elevated above other issues and resolved in a timely manner.* The CM needs to make it clear to the owner that anyone on the CM staff observing a safety hazard will bring the issue to the contractor for corrective action. Should the CM encounter an immediately dangerous situation, the CM must be empowered to suspend work on behalf of the owner, who will be ultimately responsible for the suspension.

In order to coordinate and monitor contractor safety efforts effectively, a separate safety staff should be created within the construction management team. The safety staff should be composed of safety professionals with project-specific experience and knowledge of:

- Federal, state, county, and local safety regulations
- Building Officials and Code Administrators (BOCA) and National Fire Protection Association (NFPA) codes
- American National Standards Institute (ANSI) standards
- Occupational Safety and Health Administration (OSHA) regulations
- Environmental Protection Agency (EPA), Distributed Energy Resource (DER), and other environmental regulations
- Hazard communications requirements
- Construction operations, specifications, and drawings
- Labor relations

DESIGN PHASE

The CM's safety coordinator meets with the design team to achieve an understanding of the scope for the project. At this time, the safety coordinator can be provided the opportunity to review drawings and discuss specific elements of the project, to determine potential safety hazards that may exist once the project is begun. "Design for construction safety" is a term used to integrate safety during the design of a project. For example, appropriate window elevations and parapet wall heights can help to prevent construction accidents.

PROCUREMENT PHASE

The CM's safety representative and/or client safety representative can be given the opportunity to address potential bidders at the pre-bid conference. Safety requirements pertaining to the contract are highlighted during this conference.

CONSTRUCTION PHASE

At the risk of sounding obvious, it is during the construction phase that most accidents occur. Therefore, it is vital to have "what-if" scenarios ready to go in case of a mishap. For example, the CM's safety representative should contact local authorities prior to the bid to determine the availability of ambulance services, emergency response, police, and fire units.

The CM should review the contractor's safety-related submittals to determine if the requirements of the contract specifications have been met. This review is not intended to be all-encompassing or to anticipate every job site hazard. The contract should indicate, however, that no work can begin until the safety program is approved. For some projects, it may be appropriate to allow a two-stage safety program submittal: one covering the initial 90 days for mobilization, and the second covering the remainder of the contract. The contractor's submitted program is the central element for safety compliance by the contractor and his or her subcontractors.

It is also important for the CM to develop a safety program for the CM's own employees on the job site. Construction management employees could be exposed to many of the same hazards as contractor personnel. Therefore, it is good practice to provide training for the CM's personnel.

The CM should monitor the contractor's daily construction activities and notify the contractor in writing—with copies to the owner—of any deficiencies or imminent hazards. The CM then follows up with the contractor to determine if corrective measures have been taken.

If the contractor fails to correct an unsafe condition, the CM immediately notifies the owner. The owner then notifies the contractor through the CM that the unsafe condition must be corrected, or the work will be stopped until it is.

The CM should provide monthly reports to the owner about the safety status of the program and of accident frequency and severity. Any accidents should be compared to national averages so everyone can see how the job's safety record measures up. This could indicate a shortfall or that the job site is safer than most others.

Program Management

At first blush, program management may seem similar to project management, but there are important differences. The main differences are in the size, complexity, and scope of the projects; the level of management and decision making; and the concurrency and magnitude of activity.

In most cases, a program manager manages *large capital programs*, sometimes with facilities in different locations. Often, the PM will be asked to manage or contract for activities such as securing financing, leading public relations and legislative initiatives, operating and maintaining the completed facility, and facilitating or purchasing a variety of products or services.

Where does the CM fit in? In some cases, the CM will take on the role of a program manager because CMs, by their training and experience, possess the knowledge, skills, and abilities needed for effective program management. Program management may be considered an expansion of traditional construction management services.

In his role as program manager of a large project, the CM will operate in a manner similar to that of a project manager, as we've already discussed. The only difference, again, is the scope and size of the project.

Sustainability

Sustainability has become today's watchword of construction, and it is not a fad. By all accounts, sustainability is here to stay, and CMs are on the forefront of the movement.

PRE-DESIGN PHASE

The CM's main job is to work with the owner to establish sustainability goals and objectives. A major decision is whether or not to register the project with the U.S. Green Building Council (USGBC), a non-profit organization devoted to shifting the building industry towards sustainability by providing information and standards on how sustainable buildings are designed, built, and operated. You might know the group for its development of LEED®, the Leadership in Energy and Environmental Design rating system.

During design contract development, the CM will work with the owner's legal counsel to recommend sustainability language and goals. After that, the CM will develop the Project Procedures Manual, which, like the contract, will include goals and objectives about sustainability.

DESIGN PHASE

The owner and the CM should agree on the scope and number of sustainability reviews required. The CM should establish and regularly review sustainability goals, LEED standards, and/or GREEN GLOBE standards targeted for achievement on the project.

SUSTAINABLE GLOSSARY

USGBC: The U.S. Green Building Council is a nonprofit organization devoted to shifting the building industry towards sustainability by providing information and standards on how sustainable buildings are designed, built, and operated. The USGBC is best known for the development of the Leadership in Energy and Environmental Design (LEED) rating system and Greenbuild, a green building conference.

LEED: The Leadership in Energy and Environmental Design (LEED) Green Building Rating System encourages and accelerates global adoption of sustainable green building and development practices through the creation and implementation of universally understood and accepted tools and performance criteria.

Green Building Initiative (GBI) Green Globe: Green Globe is a green management tool that includes an assessment protocol, a rating system, and a guide for integrating environmentally friendly design into both new and existing commercial buildings.

Sustainable: The condition of being able to meet the needs of present generations without compromising resources for future generations.

Building commissioning (Cx): The start-up phase of a new or remodeled building. This phase includes testing and fine-tuning of the HVAC and other systems to ensure proper functioning and adherence to design criteria. Commissioning also includes preparation of the system operation manuals and instruction of the building maintenance personnel.

Life cycle: The consecutive, interlinked stages of a product, beginning with raw materials acquisition and manufacture and continuing with its fabrication, manufacture, construction, and use, and concluding with any of a variety of recovery, recycling, or waste management options.

PROCUREMENT PHASE

Projects requiring certification with the USGBC or GBI should include appropriate requirements within the bid documents. If the owner and designer choose NOT to have the project formally registered with the USGBC or GBI, but intend for it to be a LEED or Green Globe equivalent project, then those requirements must be included in the bid documents. The bid documents should also include the sustainability requirement of bidders.

CONSTRUCTION PHASE

As in the procurement phase, all activities in the construction phase must adhere to sustainability requirements set forth by whichever group's rules are being followed. CMs should conduct separate conferences to discuss sustainability and set guidelines for recordkeeping and reporting as they relate to sustainability.

POSTCONSTRUCTION PHASE

It is the CM's responsibility to ensure that contractors have provided all documentation necessary for certification and/or required by the contract, and that the contractor and CM assigned responsibilities have been met. The Green Globe or LEED application will be submitted by the party designated as the agent for the project. This could be the CM, GC, designer, or the project's sustainability advocate.

Only when all sustainability requirements have been met and verified can the CM recommend that the owner sign off on the project.

Risk Management

All business ventures are risky, and the goal of risk management is to understand the risks involved and mitigate and manage them, if possible. One of the CM's jobs is to review project contracts for potential risks and liabilities, and to review legal requirements in the contracts to determine the potential impact of these risks. He or she then develops a plan to address these potential impacts.

The first step in risk management strategy is acknowledging that the potential for risk consequences cannot be completely eliminated but can be mitigated. Large construction projects are commonly faced with all types of risk events such as bad weather, lack of resources, unanticipated environmental factors, and community pressures. These adverse effects normally result in increased costs, resequencing of construction activities, and delays that may interfere with project delivery.

During each phase of a project, risk management should be discussed so that everyone can focus on identifying the risks that can be managed and on planning for those that may not be easily mitigated.

A Risk Management Plan should be developed and include:

Risk identification is the process of evaluating the project and recognizing the possible risks that could impact the project. It's the CM's job to organize teams to indentify risks as early as possible and to continue the identification process during all project phases.

Risk analysis provides the project team and stakeholders with a structured assessment of the potential damage each risk could do to the project. This allows the team to focus on risks that are considered to have the most likely chance of occurrence and the greatest potential project impact. The CM is instrumental in ensuring that once the risks are indentified and analyzed, the appropriate team member is assigned to keep track of the risk as the project progresses.

Risk management is, quite simply, the mitigation or management of activities that have gone wrong. The goal is to minimize the impact, with the understanding that complete mitigation may not be possible. Risk management contains four elements:

Communications and reporting: The CM holds team meetings to communicate risks and opportunities stemming from those risks.

Tracking: Each risk is followed, and potential responses are adjusted. Some risks may be retired based on the passage of time or because a specific event has been completed. The CM's job is to communicate this situation to others.

Mitigation: Possibly the most important part of risk management, mitigation is the primary responsibility of one person or group. This entity should produce a mitigation plan, which, hopefully, will reduce the impact.

Resolution: As the mitigation plan is put in place, risks come to a resolution. The risk can be avoided or eliminated, mitigated, transferred or deferred, or become a reality—a result that you don't want. As risks come to resolution, the risk data is updated to include the results. Lessons learned should be included in reports, to help identify and prevent future risks.

Building Information Modeling

Arguably the most important recent technology change in construction management is the advent of building information modeling, or BIM. BIM allows a project's entire physical functional characteristics to be displayed on a computer screen. It allows changes to be made quickly and efficiently, and is easily shared among others as these changes occur. Some models can be so precise that they can show if a swinging door will interfere with another door. In addition, not only are models shown in 3D, but they can be shown in *4D*, with time being the fourth dimension.

Relatively speaking, BIM is new on the scene, and its impact and benefits are still being discovered. Many people suggest that as BIM develops and becomes the norm for construction, the CM, as the central information hub for the project, will be the best choice to manage the high payback that BIM offers.

"Seeing" the building before it is built: one of the design benefits of building information modeling. IMAGES COURTESY OF M.ARCH ARCHITECTS.

Because BIM is evolving, the CM's relationship to this exciting technology is ever changing. To remain up to date, the CM must:

- Stay current and remain educated on the BIM process.
- Educate the owner and the project team on the benefits, features, limitations, and the implementation process for BIM.
- Continue a leadership role in the project delivery process, with appropriate application of BIM.

Building information modeling can provide a realistic view of how a project will "feel" to users when completed, as well as how it will fit into its environment. This example is the Southeast Louisiana Veterans Health Care Center, owned by the U.S. Department of Veterans Affairs. A/E: Studio NOVA, a Joint-Venture of NBBJ of Columbus, Eskew+Dumez+Ripple, and Rozas-Ward Architects. CM: Clark/ McCarthy Health Care Partners in association with Woodward Design+Build and Landis Construction. U.S. DEPARTMENT OF VETERANS AFFAIRS, SOUTHEAST LOUISIANA VETERANS HEALTH CARE SYSTEM.

BUILDING INFORMATION MANAGEMENT GLOSSARY

3D parametric modeling: 3D modeling is a superior design environment when compared to traditional 2D CAD. 3D modeling applications have the ability to capture design intent parametrically, which facilitates model creation and editing and therefore reduces the likelihood of coordination errors. Although preparation of the 3D model may be a significant part of most BIM efforts, a model alone does not constitute BIM. The 3D model, however, is a great tool for visualization of the design to benefit the project team and the project.

Engineering analysis: At the core of BIM lies a digital database in which objects, spaces, and facility characteristics are defined and stored. These characteristics make it possible to use BIM as a virtual representation of a physical facility, a representation that is capable of supporting qualitative and quantitative analyses. These BIM-enabled analyses, whether for structural, energy consumption, daylight analysis, or a number of other performance simulations, can significantly enhance the efficiency and efficacy of the design, planning, and building processes.

Clash detection: Since the 3D model represents virtual true space, a BIM process known as "clash detection" can be utilized to check for interferences by searching for intersecting volumes. It is often the case that a third-party application is used not only to clash a single model but to combine and clash multiple models from disparate sources in the same environment. An example of clash detection is learning that a door may hit a lighting fixture when it is ajar.

4D Schedules: A 4D BIM scheduling application can dynamically link the project critical path method (CPM) schedule activities to 3D objects in the BIM model. This allows for a graphically rich and animated representation of the planned construction sequence set against time. 4D schedules are a powerful tool for phasing, coordinating, and communicating planned work to a variety of audiences, including project stakeholders and those directly responsible for executing the work. These schedules also support simulated what-if scenarios.

5D cost management capabilities: Every element in the BIM model can be attributed to what it will actually represent in terms of resources and respective costs. This capability will allow a parametric and dynamic quantity take-off for bills of materials, which will result in a more accurate estimate and therefore less time spent by the estimators on the quantity take-off and more time spent on performing cost estimate analysis.

BIM integrator: The BIM integrator is a role needed when BIM is implemented with certain delivery methods—in particular, the traditional design-bid-build. The BIM integrator role can be assigned to the architect/engineer, builder, CM, or another independent party. The main responsibilities for the BIM integrator role are:

- Ensure a smooth transition of the model from the design to the construction phase.
- Maintain a central model at all times, and incorporate the latest available information from multiple project participants.

- Bring new project team members, (subs, vendors, etc.) up to speed on the BIM project objectives and current model status.
- Ensure a complete and thorough transition of the model from the construction phase to the owner.
- Verify, in all the project phases, that the model meets the owner's BIM requirements and the project BIM specifications.
- Ensure interoperability between models on those projects where a multiple-model approach is unavoidable.

From: CMAA's 2010 Standards of Practice

PRE-DESIGN PHASE

The CM must work with the owner to establish the goals and objectives of using BIM. An owner may not be familiar with the benefits of BIM, so the CM must make sure that the owner understands BIM's advantages and any shortcomings. Once the owner decides that BIM is the way to go, the CM must work with the design team, draw on their BIM experience, and learn their capabilities in using the software.

DESIGN PHASE

The CM will be in charge of a "BIM kickoff" meeting so that team members can develop and feel comfortable with BIM standards for the project. The CM and designer will use the BIM model to perform multiple design reviews, and use the modeling software to show the owner and others what the project will look like upon completion. This is an important step because it not only gets buy-in from others, but also allows everyone to see the completed project—which can reduce changes as the project progresses. During this viewing, the CM and others will learn about clashes (one door hitting another door, for instance, or telecom cable conduit paths blocked by heating ducts) and discuss how to mitigate them.

PROCUREMENT PHASE

BIM greatly changes the procurement process because bidders can see the project and understand their assignments very clearly. A lot of the ambiguity of pre-BIM days is now gone, which means that bidders can more closely hone their bids.

It's important for the CM to include BIM requirements in all contracts.

CONSTRUCTION PHASE

The CM will transition the BIM model to the construction site, making sure that all parties have access to the virtual model so they can compare their work to that of the proposal. It's important that the CM communicate the software model to workers upstream and downstream. To this end, the CM should encourage the use of the BIM model to produce shop drawings.

BIM's advantages really shine when reviewing and pricing change orders, because it allows visualization of these changes, together with an immediate analysis of their costs.

POSTCONSTRUCTION PHASE

When the job is over, the CM should make sure that the BIM model with all its changes is transferred to the owner, where it can be used for facilities management and maintenance. The CM needs to work closely with the owner's facility management team and the project team to define what the facility management team wants and needs from the model, and in what format the information will be most useful.

A BIM model can prove invaluable to an owner—not only for management and maintenance, but if and when future structural changes are desired.

4 Construction Managers' Place in the World

CMs PLAY A VITAL ROLE IN THE WORLD. Their job is to improve people's lives through the structures they help build. This includes public infrastructure projects such as bridges, tunnels, and roads, as well as private projects such as commercial buildings.

By carefully managing building factors such as costs, safety, timing, environmental impact, and sustainability, CMs not only add to the world's asset base in an efficient manner but also make certain these assets are the best that can be built under a particular situation and set of circumstances.

The effect of construction management is not limited to certain countries or regions but is felt worldwide because of globalization.

Globalization

Like almost every other business, construction management is going global. It is no longer constrained by physical borders. Construction companies that once got all of their revenues from domestic projects, now find that the lion's share of their work may be coming from outside the U.S. American CMs are in great demand overseas because of their expertise and the unique way that U.S.-trained CMs handle projects. In many ways, the U.S. is still the benchmark for how construction management is done.

However, American CMs are less inclined to work overseas than CMs from other countries, mainly because U.S. tax laws often make it less profitable for an individual to work outside the United States. Also, American CMs are often hindered by their inability to speak several languages. Both of these issues can be mitigated, through careful tax planning and the plethora of language programs available to those willing to put in the effort.

For those CMs who choose to work overseas, the rewards can be great. For one thing, American CMs working overseas will not only learn about cultures new to them, but also return with a new understanding of how construction methods and practices are performed elsewhere. Some of these ideas can be used on domestic projects, but also in those countries where the CM might work next. The expat CM will also have the opportunity to work on projects that would be unusual in the U.S., such as those involving deserts, Arctic-like environments, or jungles. These kinds of jobs may end up being once-in-a-lifetime opportunities and are both exciting and enlightening.

Working overseas also can be rewarding for those who want to move up in their company's hierarchy, as foreign service shows management their willingness to "go the extra mile."

CMs at work addressing infrastructure needs around the globe. Here, the Changi Water Reclamation Plant in Singapore. PHOTOS COURTESY OF SINGAPORE PUBLIC UTILITIES BOARD AND CH2M HILL.

Working Outside of the United States

BILL VAN WAGENEN, CCM, FCMAA

Senior VP, Construction and Program Management, Transportation Business Group

CH2M HILL

Describe your overseas experience.

› I have been doing CM and related international work since the early 1980s, primarily in the Middle East, Asia, and Europe. Those are the three areas that I have worked. My biggest program was in Korea, which is an ongoing program. I think it is still officially listed as a 10-billion-dollar program. The true numbers are going to be higher. It is to build out an expanded military base for U.S. forces south of Seoul. The purpose is twofold: One, to consolidate U.S. forces on the Korean Peninsula that are spread out across a lot of smaller sites into a more central, bigger location. Second, to vacate the base that the U.S. military has occupied in the center of Seoul since the Korean War and to return all of that land to the government of Korea and the city of Seoul for them to utilize.

The base is going to be building about 30 million square feet of new facilities. It is going to be a full military base, like a small city that will house upwards of 40,000 military personnel, their families, and support personnel. Currently, I think the estimated completion is around 2015.

What challenges do you face doing a project like this in Korea, as opposed to doing it in the United States?

› Certainly, there are very significant cultural differences, both in terms of culture and in terms of business culture and practices. Second, trying to manage two clients is a real challenge. I'm sure that is not unique; anywhere you have multiple clients there would be challenges, but here particularly because of the cultural differences. Third, this is, I think, the first real application of program management. I know program management and construction management are a little bit different, but in CMAA's world we see a lot of synergies between the two. This is the first real application of program management in Korea, particularly for the Korean government, so there is a lack of understanding of what program management really is all about.

How about workforce issues?

› In terms of our staff, we have a staff there of around 180 now, of which probably 70 or 80 percent is Korean. Because there is not a lot of familiarity with program management, it has been a challenge finding Korean staff, particularly at more senior management levels, to provide an overall PM/CM leadership. Koreans are technically very sound, but the concepts of leadership and management, at least the way we practice them in the U.S., are not as established and developed. You need to have the balance of bringing in the expatriate American leadership and management expertise in PM/CM combined with the local knowledge and technical capabilities the Koreans have. In Korea, also, CM is a somewhat more rigid, ruled, and regulated directed practice. The Korean government has various regulations defining what CM is, how it should be practiced. The main point is that it tends to be more regulatory driven, as a practice, than, certainly, it is here in the U.S.

A national treasure, Teatro Colon in Buenos Aires, Argentina, was carefully restored and modernized under the management of Grupo SYASA. PHOTOS BY ADRIAN PEREZ FOR GRUPO SYASA.

How about logistics and resources? Is that a challenge there?

› No, not really. The Korean construction industry is pretty well developed; I would say nothing is significantly different than in the U.S. They have a lot of capability in Korea for fabricating and manufacturing sourcing materials and so forth, and delivering. They have a fairly well-trained construction workforce in Korea. I think if you were to look at industry estimating data, their productivity level, as compared to the U.S., is lower. In essence, you would need higher labor hours, but that is offset to some extent by lower labor costs. That is a sensitive issue when we have raised that in our program. Talking to the Koreans when we have done baseline estimating, they have sometimes reacted a bit negatively. If you look across global industry estimating standards, they will tell you productivity is a bit lower in Korea than it is in the U.S.

What global CM trends are you seeing?

› First, I think there is a growing recognition of PM/CM around the world, particularly in developing countries. Second, along with the growing recognition, is a real desire to learn. I think, in that respect, the U.S. continues to be recognized as the global leader in PM/CM in the world. Although there are quite a number of countries that are very proficient in CM, including European countries, South Africa, and Australia, I think the U.S., in general, is still seen as a global leader. There is recognition, and the U.S. is seen as a leader, so there is a desire to learn from [the U.S.] and obtain practices that the U.S. promotes in terms of PM/CM. I see more recognition of and desire for formal certification around the world. Some of that is, quite simply, a way for people in other countries, particularly developing countries, to gain a recognized certification that they could then use to better market themselves.

I think it is becoming more competitive for Americans to work overseas. This means that as you are working overseas in these countries, you really have to be focused on developing local partnerships, maximizing your use of local partners, subs, and resources there to deliver your services and limit, to the extent that you can, the number of expats. I think there is beginning to be more recognition of risk and effective risk management internationally.

Challenges of Globalization

American CMs working overseas face challenges they don't see at home. Aside from language and cultural issues, CMs working in remote areas of the world may encounter logistics and materials issues. The U.S. and most industrialized countries enjoy a robust transportation infrastructure. It's a fairly simple matter, relatively speaking, to order materials and have them delivered when you need them. In some other areas, this may not be as easy to accomplish. The same goes for heavy equipment. The American CM in remote areas may find that getting the right equipment at the right time is a daunting task. It may require more planning and effort than the CM is used to.

Being a CM overseas may also require a different attitude toward the workforce. Such seemingly simple tasks as hiring and scheduling may require extra planning, as employment laws and customs vary from country to country and even from region to region within the same country.

A U.S. CM will have to work a little harder at building trust with his or her staff. After all, he or she is the outsider and is unfamiliar with how things are usually done. It requires even greater emphasis on people skills than if the job were in the U.S. CMs usually hire a coterie of interpreters and others who can help them understand the nuances of job sites, how orders are given, and what can be expected from others in terms of their behavior.

The National Library of Latvia, one of the most complex and ambitious construction projects ever undertaken in that nation. The main role for the project managers, Hill International, was to improve buildability. "We were liaising with the architects and structural engineers to make sure we were coming up with a design that could be built in Latvia," explains the project leader. "It's all very well having a complicated design that could be built easily in New York but that isn't helpful here." Hill International provided CM services. PHOTO COURTESY OF HILL INTERNATIONAL, INC. © 2011.

Baku "Flame Towers" in Baku, Azerbaijan. CM by Hill International, which provided this photo. PHOTO COURTESY OF HILL INTERNATIONAL, INC. © 2011.

CMs who have only worked in the U.S. are used to working aboveboard with permitting and licensing agencies, both on the state and local level. Except for isolated instances, government agencies are fair and honest. Obtaining building permits, for instance, may be a lengthy or exasperating experience, but it will be forthright. In some other countries, however, bribery and payoffs are commonplace, and CMs should be ready to handle these issues and comply with the U.S. government's laws as well as the values and regulations of their own companies. In some cases, U.S. companies have gotten into hot water because they paid bribes. In other cases, they have given up or not even bid on projects because the corruption environment was too severe. One of the jobs of the CM is to assess these hurdles before the company considers bidding on a project.

Cultural Differences in CM

JIM MITCHELL, CCM

Project Manager

AECOM

Tell me about AECOM and your work.

› AECOM has 50,000 employees around the world. We are a multidisciplined A/E/C firm, and we're in all markets and geographies, including working in over 100 countries around the world. We do design work, architecture, engineering, program construction management, and project management. One of our more recent acquisitions was Tishman Construction in July/August. We offer a fully integrated solution, from advance planning to own-and-operate and everything in between. I run the PM/CM business for our facilities-related or vertical work. I'm not really involved in the horizontal transportation or things like that; I'm on the building side. We are in the process of putting together our construction services, which my group will be a part of along with Tishman and Davis Langdon, which was another recent acquisition. The group I lead is predominantly program and construction managers around the globe. Combined with Tishman and other regions, we have about 4,000 people out of the 50,000.

Based on your experience, how is CM practiced overseas, compared to the United States?

› The CM is similar; the big difference occurs in program and project management. There is very little program management the way it is defined, practiced, and done in the U.S. There is lots of project management, single-entity projects. The difference is that the quantity surveyor or the cost manager gets involved very early elsewhere in the world and works as the trusted advisor to the owner, helps the architect with conceptual estimating, the initial estimates, and the design scope to budget.

Why do you think that is?

› It's just the European way of doing it. That industry has grown up with quantity surveying.

Do you think that it's better, worse, or neutral?

› I think it is neutral; I do not see a great need. When you compare the way we do it in the U.S. versus there, why have different people doing it? Program management, or those front-end services, is typically part of what the project manager or program manager provides the clients under the CMAA definition. I know clients in the Middle

East that do not want anything to do with the quality management system (QMS) way of delivering projects, they only want it done the way the Americans do it. We are starting to see a lot of that; they say, "We want Americans to run our projects."

Why do you think some others want to have it done the "American way"?

› I think it is because as the projects get larger and larger, they are not single operations anymore. They are very complex. The problem America has is that there is a great need for Americans to work overseas, but most Americans do not want to work internationally. They want to stay right here. Getting them to move from Massachusetts to Virginia, or even Boston to Worcester, is a nightmare.

The problem often gets back to money. Is an American project manager cost competitive with an Australian or Brit? The answer is "no" because of the IRS. The IRS says to the expat, "You only get to exclude the first $80,000, $90,000, or whatever that number is, as tax free; everything else is taxed." To a Kiwi, Aussie, or someone else, it is all free. The differential to get an expat from another country to that location—say, the Middle East—is substantially less. A U.S. person is going to say, "Wait, you need to give me tax equalization."

What other challenges are there for a company or CM working overseas? What about materials or logistics?

› I think, short of being in a third-world country, if you're in the Middle East, Abu Dhabi, Saudi, or Dubai, you don't have those problems. What you have is somebody coming from the U.S., looking at it, and saying, "Why are they doing it this way?" You have some cultural issues, for example, the terminology that we use in the U.S. We call it a bid. If you say, "What is the bid date?" they are going to look at you and say, "What are you talking about?" If you say, "When is the tender date?" now you are speaking internationally. Even within my own company, there is a difference if I refer to a CM to somebody in Australia or in the U.K. If I say CM to you, what do you think? Construction manager or construction management. They think *cost management*. There are differences just between the common acronyms that we use in the industry.

In the U.S., usually if you do A, B, and C, you will get D, whether it has to do with permitting or licensing. No one, in general, is shaking you down. But you hear stories about that kind of problem occurring in other countries. What is your experience?

› I hear both, depending on what country you are in. I can take you to cities in this country where, if you want a permit, you have to go to 10 different permit payment locations. In some of the African nations, it is a little bit more prevalent than others. In Australia, it is just like doing it here.

I haven't personally seen shakedowns. We are working in the Middle East, and especially in the monarchies, generally their permitting authorities are all relatives. But, no, I have not experienced illegal activity. For a U.S. company working overseas, it is key that they understand the Foreign Corrupt Practices Act. It doesn't matter that they might be housed in Germany. If they are tied to a U.S. company, the law applies. There are also the ITAR [International Traffic in Arms Regulations] regulations that companies working internationally need to understand. These are regulations that have to do with supplies going to different countries. For example, as silly as it sounds, if John Deere wants to send a combine to Russia, they have to clear that with the Feds. They have to demonstrate that the combine cannot be used or converted into a weapon of war. There is a pretty extensive barrier to entry for companies that haven't worked overseas to start working internationally.

What about language issues? Americans are pretty bad at learning other languages; we tend not to do it. But people in other countries may speak two, three, or more. Is that a factor on the job site?

› It is and it isn't. In areas where it is, we generally get interpreters to be with our project leadership. The other thing that is a challenge for U.S. companies and individuals working internationally is, short of the U.S., everything else is metric. For our engineers, architects, and constructors, they have to learn something might be three meters long instead of 10 feet.

Where does the U.S. stand on BIM compared to other countries?

› I see that Europe, England, Australia, and some other places are a little bit further ahead, but I think we are rapidly closing the gap. When and if the U.S. government ever defines a standard, it will leapfrog. People think BIM is going to be the panacea of the construction industry; it is not going to be, it is not going to build the building. It may solve some clash detection, and help improve the design. At the end of the day, your design needs to be great. If you have an uneducated and untrained workforce, you are going to have the same poor construction, and it does not matter what the drawings look like.

Any last words on the global scene and CM?

› I think it is a great opportunity. For AECOM working internationally, this is the second year that we have had in excess of 50 percent of our revenue coming from non-U.S. sources, yet we are a U.S. company. I think young or old people coming out of school, wanting to see the world and really make themselves highly marketable, need to be willing to do an international assignment. They need to be willing to do it in China or India, as well.

If you look at the GDP projections for the U.S. versus some of the other emerging markets, we are stagnant for the next five or ten years. The thing about working in China is that you work at Mach 4 for about four years, and then you are burned out. If you want to do big projects, that is where they are being done; there or the Middle East.

Infrastructure

The term infrastructure refers to the many structures upon which a society depends. These include roads, bridges, tunnels, water supply sewers, power grids, and telecommunications. It also includes buildings such as schools, libraries, government buildings, and hospitals.

For many CMs, infrastructure is where the action is. They are excited about building large, complex projects because of the challenges they present. It stretches the CMs and calls on them to employ all of their expertise and experience.

These CMs are also attracted by infrastructure projects because they serve the public. Many CMS are proud to be able to drive over a bridge or past a school and say, "I had a hand in building that."

Unlike a commercial building project, large government-sponsored infrastructure jobs present a unique set of challenges.

Renovations and modernization of H. D. Cooke Elementary School, Washington, DC. CM by Gilbane Construction. PHOTOS BY JEFF KATZ/2010.

Special Challenges of Building Schools

TOM ROGÉR

Program Director Rochester Schools Modernization Program

GILBANE, INC.

Tell me about Gilbane's work in the area of school construction.

› The company tries to concentrate its experience by market segment; we call them centers of excellence. People tend to gravitate towards one of these different market segments in terms of their project experience. We create these centers of excellence so we can share that experience as people move around within the various regions and get involved in different projects. I have been the head of the center of excellence for K–12 schools ever since we started the concept of centers of excellence. We have a monthly call of all the different people in the company who actively participate as part of their center of excellence. We share war stories, talk about upcoming project opportunities, and try to take full advantage of what people are doing. For big companies, like Gilbane, one of the downsides is that we are all working on our individual projects. We have the benefit of our personal experience, but you can only get so much from that experience. We run into all kinds of problems, and it's nice if you are not trying to reinvent the wheel each time you confront these problems.

What are some of the challenges that are unique to school construction?

› Funding availability is the big challenge. School projects, and other public projects, are more complicated in terms of their funding stream. It varies by state and even by school district. There are 16,000 school districts in the country. When you look at the larger districts—we tend to concentrate on larger projects and larger districts—they have to cobble together their funding for capital programs. It runs the full gamut from requiring voter approval to bond referenda. Sometimes, state legislative approval of capital funding programs is required. I have done most of my work in Connecticut, Massachusetts, and New York. They all have fairly generous state programs for capital funding. There is a whole long process as to how a district qualifies for capital funding, everything from individual project filings and approvals to getting X amount of dollars allocated to a district. They are all different. Where the states have these programs, they tend to be the more active areas for school construction. A lot of the states do not have state programs. Probably 30 percent of the states—Tennessee is a good example—have no state funding programs, so all the capital funding has to occur at the district level. The wealthy suburban districts keep pounding out new high schools and elementary schools, while the poor urban districts are in terrible shape.

As a CM, how do you get involved in that process? Is it different from getting involved with a private owner who, say, wants to build a building downtown?

› It is. Where possible, we try to assist the districts in coming up with estimates if they need to go for bond referendum funding, and a lot of districts have to work under that system. If they are going to go for a $100 or $200 million bond referendum for a project or a group of projects, we try to help them come up with the right estimate so they actually get the money and do the project, and so they don't end up with a problem of an overrun budget. A lot of states have rules about how much we can be involved in terms of helping them politically campaign, so we function more as a technical consultant in that phase of the program. Once they have project funding, we want to be on board as early as possible to help them manage the budget and the delivery process. We have a variety of skills that we can bring to the owner early on. A lot of times, we do it as a freebie to help the owner move into the real world until they can actually hire us and have funding to hire us as a consultant or CM.

What about the actual construction itself? Are there any challenges that you don't see in other projects?

› Yes, because of the fact that, number one, they are school buildings. The type of building and the type of standards you have to meet, generally brings a higher level of oversight. Particularly if there is state funding involved, oftentimes the plans have to be approved at the state level. You may have a double level of approval involved; not only do you have to satisfy the user, but you have to then take it to the local building officials to get their approval. Then maybe again to the state officials to get their approval because they could be providing 90 or 100 percent of the funding. There tends to be a longer approval cycle.

Also, many times buildings are being occupied during construction. You have, not only the notion of not affecting the existing operations, but a whole level of security to be concerned about. Many states have state regulations about what type of work can be done in the building if it is occupied. For instance, you cannot do asbestos or hazardous materials removal if the building is occupied by elementary kids. You cannot do roofing work; there is a whole list of rules you have to comply with.

Is there something particularly more fun or challenging about doing schools? What is the kick for you?

› From hospitals to pharmaceutical facilities—you name it, and I've done it. I personally like schools because I feel if you do a good job, you can positively affect people's lives. That sounds trite. Particularly when you go into an urban situation, where you are looking at school buildings that are way past their useful life, and kids and teachers who are literally suffering in these buildings. To be able to rebuild that space and provide a much nicer environment is rewarding. I have actually had teachers come into finished buildings and cry at how pleasant the space is. That is unusually rewarding. There is a big community benefit to rebuilding school buildings.

What trends do you see in school construction?

› Over the last 10 years, I have seen more attention being paid to the effectiveness of the whole educational system. I have always been told that good education is three P's: people, program, and place. The concentration has largely been on the people and the program, with things like rating teachers and school programs. That has been the concentration, but the "place" part of it has come into the equation. I have seen now that people are beginning, largely fueled by people like myself and the architectural community, to say, "Certainly, you can have good educational programs in lousy buildings, but you can attract better people and provide better programs if you have good buildings." People are not just building space; they are trying to build quality space. Particularly, they are beginning to understand that these are long-term investments. These sorts of projects, a lot of times, are knee-jerk type reactions. Now that people are rebuilding space built back in the 1950s and 1960s, when the baby boomers went through, they are looking at the space and saying, "This is awful, why did we do this?" Now that they have to make this investment to either renovate the space or rebuild it, they are in fact making a more reasoned approach to the type of space that they're building. They're seeing it not as just creating space, but really trying to create more of an institutional building like they had in the 1920s, when they built the good buildings. It is funny how it has come back full circle.

Washington-Lee High School in Arlington, VA. A three-and-a-half-year project involved a complete rebuilding: Some 250,000 square feet of existing construction was demolished, and a new, 330,000-square-foot facility built in its place. The project earned LEED Gold certification. The project team implemented a phased schedule that enabled the entire project to be completed without interfering with the ongoing operation of the school. PHOTOS COURTESY OF MCDONOUGH BOLYARD PECK, INC. AND GRIMM + PARKER ARCHITECTS.

For example, building a school is more than just a matter of placing brick and mortar. CMs may find themselves involved from the very beginning, helping local jurisdictions present financing options to their government agencies. This provides an important service to school districts, for example, because they don't have expertise in this area. Their job is to educate students and not build schools. As more and more jurisdictions find their budgets cut—not just for schools—CMs are seeing that their skills to come in early on a project to help determine a realistic budget and time line are in demand.

CMs also find that building infrastructure projects presents logistical issues not found in other projects. Because current infrastructure is always being used—roads, schools, hospitals—they must muster all of their expertise in keeping these services active while construction goes on. Unlike building a skyscraper on a empty lot—where there are no occupants—water must still flow, and hospital patients must still be treated, while these particular structures are being renovated or added to.

Understanding the Differences

CHUCK DAHILL

President

CM/PM division of ARCADIS

You are involved in infrastructure work all over the world. What are some of the trends that you see?

› The biggest thing that is going to change on the CM side is going to be less on the CM, construction side, and more on the upfront planning, preparation side. That is because of the different delivery methods that are being forecast and used today. It is going to take a lot of the responsibility away from the owners, whom we typically represent, and put it more on the contractor or design/build team. I think there will be less emphasis on the construction phase as we go forward.

Why do you think that is?

› Because of the way contracts are written. Under the current economic status of funding, I think you are going to find more design/build with funding involved into it where the contractor and design/build team will have more responsibility for longer-term operation and maintenance, rather than just building it and turning it over to someone.

As far as financing is concerned, where do you see things going?

› I think you're going to see significantly more public private partnerships getting involved. You'll see more private sector involvement, funding, ownership, and operation of projects.

Do you think we'll see things like TIFIA (The U.S. Transportation Infrastructure Finance and Innovation Act) in other countries?

› Yes, absolutely, in the developing countries especially. I'm not deeply involved in our international business. I sit on our Global Board, but I can tell you I think in the fastest-growing parts of the world, funding is going to be a key component.

Central Weber Waste Water Treatment Plant in Ogden, UT, and Thomas P. Smith Waste Water Reclamation Facility, Tallahassee, FL. Water-related projects, both for drinking water and collection and treatment of storm- and wastewater, will be an increasingly high priority for American CMs in the years ahead.
PHOTOS © OF MWH CONSTRUCTORS, INC.

What about workforce issues?

› Early on in my career at least, if you went into the construction business, relocating was standard. You went from job to job, you moved and took your family, or you left the company and went somewhere else. We're seeing a significant change to where people are not interested in doing that; you have two-income families and changes in priorities. I think you have that impact. I think you have another impact that I'm concerned about, and that is fewer people want to get involved in our business going forward, and people are more interested in—you know, we went through the eighties and nineties where everybody wanted to be on Wall Street—now, everybody wants to work for Google. So I think the challenge that we have is making people aware of the opportunities that are in the industry and that it is exciting and new and it's interesting.

When I told my parents that I was studying engineering construction they were really disappointed because they thought I was going to be a carpenter or a framer or a plumbing guy. I said I was going to be an engineer and would be a manager in construction, and they said: "So you are going to be a general contractor and be a framer?" I said: "No, that's not what we do." So I think we need more education, explain to the population about the opportunities that are out there in the construction industry. Early in my career path, most of the guys that I talked to, to interview for jobs, were trades guys who worked their way up and became big contractors. Now, however, the management of construction is very professional and complicated, and that has a lot to do with the risk.

I think one of the biggest challenges we have is confusion over definition about what construction management is. We work really hard trying to differentiate program management from project management, from construction management.

The Challenge of Keeping the Water Running

CATHY GERALI

District Manager, Metro Wastewater Reclamation District

Denver, Colorado

Describe your upgrading project.

› We are the largest wastewater facility between probably St. Louis and California. Our capacity is 220 million gallons a day. We are currently doing not only major upgrades to our existing plant to meet new regulatory requirements and permit standards, but we are also building a $475 million satellite plant. It will be our first satellite plant. It is scheduled to be online by 2015, and all of the upgrades at the existing plant have to be completed at the same time. We are really running.

Do you use outside consultants?

› We have always used outside support from consultants. Back in about 2005, roughly, we had been clicking along from about 1995 to 2005 spending maybe around $20 million a year on capital improvements. Around 2005, through our capital planning process, we knew these big projects were out there looming on the horizon. We were looking at going from about $20 million in expenditures a year to probably, in another two to three years, peaking out at over $200 million of capital expenditures. We simply knew we didn't have enough staff on board to deal with all of that. It just makes absolute sense to use outside consultants and bring in their expertise.

What is the CM's most important role, and what are some of the challenges for a CM in doing this kind of infrastructure work?

› As manager, I expect a CM to become an extension of district staff. I want them to be integrated in our process, understand our procedures, and how we approach projects. I really believe it absolutely has to be a team effort. Just as an example, we are very good at holding a club over CMs' heads, and we charge all sorts of fees if they do not meet certain requirements. I have truly tried to move to more of a cooperative situation with the CMs. Like I said, you have to be a team. You cannot just hit them up with penalties. One of the things we have done in the last couple of years is put in bonus payments. It has been hugely successful.

Bonus payments for faster work?

› Yes, to get the work completed quicker. The main reason we did it on our existing plant side is that we had some major upgrades that had to be done before we tackled the big rehab program we are doing now. The sooner those projects were up and running, it made it easier for us to move on to the next phase. We actually have one project that is going to be substantially complete, I believe, in March of 2011. It is a year-and-a-half ahead of schedule and absolutely a total success. I believe—I would have to verify this—the CM and the contractor earned about $900,000 in bonuses.

Wow, that is incentive.

› It is. I think historically the District has always been concerned that if you provide those types of incentives, contractors will rush through the work

Brightwater Marine Outfall in Shoreline, WA, is a pipeline that conveys treated wastewater from the end of a conveyance tunnel to a discharge location about one mile offshore and 600 feet deep, the deepest outfall in the United States.

and not do a good job. I absolutely did not see that. This next phase of our project that we are moving into is a rehab of around $220 million. We are doing the same thing; we are putting in bonuses for the contractors.

It sounds as if you are moving towards an Integrated Project Delivery (IPD) model, away from the blame game and towards the carrot rather than the stick.

›Exactly. It doesn't mean that we do not have our issues. I think every project does. I think going into these projects it is a much more positive approach, both from our side and the contractor's side. I think it is a huge benefit. Typically, as an owner, owners have to have a reason [for paying a bonus], especially because we are considered a special district, but basically we are using public funds. We have to be accountable for the use of that money. I think moving forward into the next 10, 20, 30 years, I truly believe that is the way owners should go, and I think contractors should look at it.

The kind of projects you are working on—wastewater, sewer, and that sort of infrastructure—do they require a CM to have a different skill set, or a different attitude, or a different kind of education?

›We look for expertise in the wastewater processes and understanding what is required to meet our discharge permits, which we cannot violate. CMs need to have specific expertise in wastewater, but my perception in managing a project, whether it is a road or a new wastewater plant, is that management skills cross all of those areas.

Typically, in our industry, we do what is called a design-bid-build. You design it; you go out and have a 100 percent design; you bid it; you have a set contract amount, and then you construct it. In design-build, those efforts are being done consecutively or concurrently. The owner's side has been a challenge for us simply because we have not had much experience doing that. When we knew we needed to start moving to these alternative deliveries, there again, we had to look to the CM world to bring in the expertise and the knowledge to help us get through these.

The reason we are doing the alternative deliveries is primarily because of schedule. We have made commitments to provide service to new entities, so we need some of these facilities online as we

The key to success for this project was a very tight schedule. Construction could only take place during extremely limited time windows, due to permits that protected migrating salmon and trout. Anticipating these limitations, Vanir Construction Management and its colleagues brought the project in more than 21 months ahead of schedule. PHOTOS BY DANNY WEBB, VANIR CONSTRUCTION MANAGEMENT, INC.

have committed to a schedule. As an example: When we knew we were going to move forward with this, we did a small plant project that was roughly for $1.5 million; we did that as an alternative delivery. Then, about eight months later, we did another project of about $22 million as alternative delivery. Now, we are jumping to this $475 million project as alternative delivery. Obviously, from a staff perspective, we are on a huge learning curve. It makes a lot more sense to be able to go out and bring in specific expertise as we need it, rather than trying to have it on staff all the time.

You have some challenges that other CMs do not have: You have to keep the water going; you have to keep the sewer going; you have to deal with political pressures and all these things. Can you summarize the greatest challenges you have being in charge of a project like this?

› Our biggest challenge right now is that we are spending a lot of money in a very strong economic downturn. We must communicate to our board and our ratepayers, who have to pay for this, and educate them on the need for it. That is, obviously, a huge part of the challenge we are facing. On the national level, if you are not in the wastewater/water business, you probably have not heard a lot about this, so some of the national organizations involved are spending a lot of time trying to educate on a national level. That includes EPA and government officials in Washington. You cannot keep throwing more and more stringent regulations at us that we have to meet, without looking at the dollar side of it. I think sometimes those two are not looked at together. There is a big movement right now that we are involved in with an organization called the National Association of Clean Water Agencies. The endeavor is called Money Matters. With utilities continually being forced to meet these regulations, I think we have gotten to a point where the improvements we are making are so miniscule. I do not believe a lot of them are really cost effective.

I think it is just the world we are in right now.

Exciting Times for Infrastructure

As the world pulls out of a recession, infrastructure has become a hot issue. During the recent downturn, many projects were put on hold, while others never even got started.

In the United States, for example, much of the infrastructure built during past decades is in dire need of replacement, especially roads, tunnels, and bridges.

The American Society of Civil Engineers has been issuing infrastructure "Report Cards" for many years, calling attention to the poor and deteriorating condition of these critical assets. The 2009 Report Card gave all American infrastructure a grade of "D" and estimated it would take an investment of $2.2 *trillion* over the next five years to address these inadequacies.

Aero-Train People Mover under construction at Washington Dulles International Airport. PHOTO PROVIDED COURTESY OF METROPOLITAN WASHINGTON AIRPORTS AUTHORITY.

From aviation to dams to transit, America's infrastructure needs are urgent and becoming more so. This plethora of overdue work presents many challenges to CMs, who are not only in short supply but find that funding is getting more difficult to obtain. Taxpayers are becoming more critical of infrastructure spending. Rightfully so, they want a bigger bang for their buck, and this puts a lot of pressure on CMs to not only budget tightly but perform their work in the most cost-effective way possible. (More on innovative financing in Chapter 5.)

CMs are also being called on to build structures in ways that were not required beforehand. For example, the threat of terrorism is reflected in buildings with reinforced walls. Buildings and bridges that were never built to withstand seismic shocks are now being specified for this capability. (For example, the San Francisco Public Utilities Commission is in the midst of a $4.6 billion program to modernize and seismically refit its huge public drinking water supply system.)

In general, building codes are becoming more stringent and present CMs with great challenges to give greater asset value while keeping prices competitive.

Lots of Infrastructure Projects Ahead

JOSEPH P. McATEE, PE, FCMAA

Executive Vice President and COO

URBAN ENGINEERS, INC.

What's your outlook for infrastructure construction as it relates to CMs?

› My specialty area, my firm, is basically public infrastructure, whether that is ports, marine, aviation, transit, highways, or bridges. I see a challenging future. I think the basic infrastructure needs of the U.S. are grossly underfunded. The political leadership has been unwilling to make the investments necessary, and that is basically because they are concerned about losing votes by raising taxes on infrastructure. I see a wave of public private partnerships (P3) coming out in the next several years, and how CMs are integrated into the P3 initiative remains to be seen. It will be a critical marriage of the folks who have the technical discipline to deliver projects and the financial partners that are going to be funding the projects. That is a little bit of the challenge. I see it being a very integral part of project delivery as it is now, and it will get a lot broader in the future because of the advent of privatization in public private partnerships utilized by the various states and municipalities. The CM will have to pick a place in the entire cycle of project delivery. Right now, CMs advocate providing the service from the beginning of scope development through the end of construction. That is not widely used in the public infrastructure area, but it is going to have to become more widely used.

Talk a little bit more about construction managers' roles.

› When you are talking about using CM more broadly and using it from the beginning of scope definition of a project, the public owners will be bringing CMs on board much earlier, to allow them to have the benefit of a professional with expertise in identifying the scope of a project and implementing skills to control the project's time cost quality. They will do this using project management skills that allow them to control what, very often, drives a lot of the cost issues, and that is time.

Time is really king in a lot of these areas. When you get into the public infrastructure area, you have a lot of folks doing planning, evaluations, permitting, and things that are not directly related to building a project, but are contributing to developing the project. Very often, these folks are not focused on the time needs of delivering the work to control the cost. A good CM, introduced into the program early, will put together a management performance system that identifies where decisions have to be made early, to allow the owner to enjoy the benefit of limiting his overall cost per program to deliver. In other words, right now, you may have a number of professionals attend a meeting, and after the meeting is over there will not be any real action items or outcomes from the meeting. Good CMs will not let that happen; they will have deliverables, action items, or the next steps that have to be taken—and identify who has the action. They will develop an overall faster schedule of the elements of the overall program that were discussed in the meeting and how they are going to go forward and achieve the next schedule items. It's very schedule-focused.

The current folks in the planning and design arena do not really think about the schedule as a hard deliverable. That is a big discriminator right now. It certainly becomes a big deliverable when you have a contractor who says that they are going to build something in a certain amount of time; then the schedule takes a front-row seat. When you're in a design, because of the variability of issues and changes that are very frequently introduced, schedule is just something you talk about and try to achieve, but it is really not driven hard by some professional.

What technologies do you think CMs will be using more? These may, in fact, be things we do not have yet. Obviously, BIM is going to be bigger.

› Yes, we are expecting BIM to be the biggest thing . . . certainly in the vertical environment. I do not see clearly how it is supposed to be integrated into the flat environment of highways, bridges, and that type of thing. Certainly, on an aviation project, where you have a runway and a terminal, the terminal will be BIM. There is no question about it. The big question right now is: How does the CM fit into a BIM project? Does the CM control the BIM program? Does the CM oversee the architect doing it? Does the builder's developer control it? That goes to whatever project delivery system you are using. If you design then build, it is one way. If it is a private-size project, it might be a different way. I think these various methodologies of using BIM are going to be somewhat germane to each different market area.

Where do you think IPD will fit in the coming years?

› I think if you put yourself in the future of public infrastructure and think about privatized projects, public private partnerships, and that type of thing, there is a chance that states would empower the entity to use IPD. There is a chance of that. I say that because right now many of the states, maybe all the states—I have not done a survey—have public procurement laws that require a low-bid environment, public bid, full transparency. There is no methodology out there now that I know of, except I believe in California, where you can have a contract, an owner, a public owner, a design professional, and a contractor at the very beginning of design. Right now, the states have to go through a process of completing

Pacific Coast Highway Traffic Congestion Relief Project, Dana Point, CA. CM was provided by the City of Dana Point, PBS&J (now Atkins) and TWNiemann, Inc. PHOTO BY BALANCE DIGITAL IMAGING.

a design and then advertising it and bidding it to the contractor. That is definitely going to change. There will definitely be a niche for IPD; there is no question about it. There are too many benefits. I see it working its way into probably a 10 to 20 percent mode of certain projects where there is flexibility.

It seems that with IPD, the role of CM becomes even more vital. Would you agree?

› I certainly would. I have had some exposure to some project brown bags, or presentations, if you will. I had one contractor, a big company who does a lot of big work. They use BIM, and I think they have been using IPD also. They see themselves as a CM; they are the at-risk builder.

One thing I think is really going to affect the marketplace going forward is globalization, such as, getting professionals from other countries, working in other countries, offshoring more design work, manufacturing a lot of materials that are integrated into U.S. projects but are fabricated abroad. Right now, there is a lot of legislation against fabricating steel abroad, although there are some exceptions. We are working on the Freedom Tower in New York City, and that has a lot of steel fabricated in Spain, as an example. The CM has to integrate his or her services with an international face. I see that happening more and more, where, if you are doing CM on a major facility in the U.S., you will have to have people that can be deployed to remote areas to do quality assurance evaluations on the materials being provided and fabricated. That is going to be more the case in the future.

CM and the Environment

Going green is not a phase or fad. It is here to stay and will impact how CMs do their work from now on.

Many owners are demanding that their projects be environmentally sound. This is important not just for the obvious societal benefits but because governments are offering tax and other incentives for projects that meet certain environmental criteria. Indeed, there also are environmental laws and regulations that must be obeyed, or fines, sanctions, and shutdowns can be imposed.

More and more, CMs are being called on to assess the environmental impact of buildings both ecologically and financially. They are also being asked to make sure that environmental regulations and laws are being followed during construction.

Sustainability is closely related to environment impact. While CMs are concerned with how their projects affect today's ecology, sustainability looks into the future effect of a project—how it will impact future generations. This was covered in Chapter 3, but the definition of sustainability bears highlighting: *Sustainability is the condition of being able to meet the needs of present generations without compromising resources for future generations.* Sustainability goes beyond environmental issues, although the two are intertwined. For example, sustainability may anticipate the direction in which buildings codes are projected to move, usually to increased stringency. CMs may also consider what an area's water needs may be in, say, 10 years, and build a structure accordingly.

Anticipating future environmental conditions is, of course, impossible. But CMs can use best practices to build structures that they believe will have less impact on future generations. One thing is certain: The more we learn about the environment, the more we realize the fragility of the ecosystem and how much humankind has affected it—often negatively. In the future, environmental rules and regulations will continue to change, based on new information and new scientific discoveries. The challenge for CMs is to remain current on these issues and adjust their practices accordingly.

It is also incumbent upon CMs to lead the way in environmental excellence in construction and to spread eco-friendly ideas to areas of the world where there now is less awareness.

Two views of the huge City Center project in Las Vegas, NV. Sustainability, particularly attention to water conservation and reuse, is shaping development plans all over the world, but especially in areas of scarce water supply such as the desert. PHOTOS COURTESY CITY CENTERLAND, LLC

Sustainability: Going beyond the Building Codes

LONNIE COPLEN
Project Manager
McKISSACK GROUP

How do you describe sustainability?

› The process of attacking a project from a sustainable construction perspective takes the best of what a CM has to offer. A great CM, we all assume, brings really good communication and coordination skills, which essentially means the ability to talk to different people about the same thing in a language they understand, and understanding the relationship of building or construction systems to the whole of the project in context of the cost and the schedule. The most effective approach to a sustainable construction project is probably also a very effective approach to good CM, and vice versa. If it is a good CM approach, it will probably work well on a sustainable project because it recognizes the relationship, or interconnectedness, or interdependence, of various building systems and the really important, critical nature of thinking about a project across engineering disciplines or across trade skill sets, or trade silos. "Silo" is a word that people like to use. A good CM approach recognizes the importance of thinking across those disciplines and trade silos in order to make the project work at the end of the day.

How would you answer people who say sustainability is a given?

› I think the vast majority of people you will speak with will maintain that, until the construction folks and CM folks live and breathe the principles of sustainability, this whole movement does not make sense because construction people won't know how to achieve the principles by way of doing the things in the field the way they need to be done. Whether you have the principles of sustainability at the forefront of your thinking as a designer, engineer, CM, or a contractor, the code is what we all set our sights to achieving. The thing to recognize about sustainability, at least as it is in our society today, is the assumption that these projects go beyond the code towards something that is, to my inclination, more energy efficient. Our codes are not there yet.

If CMs are not speaking the language of life-cycle management, shame on us for not driving the process to help our cause as well. We can all be growing our fluency around life-cycle asset management in a way that can help our clients, our industry, and help the environment to bring it back to the sustainability topic.

Challenging Work on Federal Infrastructure

JOE GRAF, PE, CCM

Describe the work you and your company are doing.

› I am the head of the professional services line for the Shaw Federal Infrastructure Group, and our clients are purely federal, and this is an area that we are putting added emphasis and focus on. We have dabbled in it in the past, but it has been hit and miss. A contract here, a contract there, but there was no real leadership behind it and therefore no cohesive strategy for making it in a legitimate business line.

What jobs, in particular, are of interest?

› We are an infrastructure-oriented group, although we do some vertical construction, so we are looking at engineering and architecture services for our clients with a predominance towards the infrastructure side of things—utilities and roads and bridges and mass transit and those kinds of projects.

What are some of the challenges CMs face working on federal infrastructure projects?

› One of the most fundamental elements of a new CM coming into the business is the ability to communicate, both in writing and orally, and to facilitate a project team to achieve its goals—and I find that to be a weakness in many new graduates, even those who come from fairly prestigious programs, because somewhere we have lost our ability to effectively write and provide oral kinds of skill sets.

Oftentimes, we get a contract at the federal level issued from the headquarters organization, but the end user is located on a military installation or some national park. When you are managing that project, you've got to understand who is holding the contract and what are the requirements of that contract. There are sometimes conflicts where the ultimate end user wants one thing, the local federal manger wants another, and the contract holder in Washington, or wherever, has a third and ultimately contractual viewpoint. So the challenges for that project manager are to recognize the varying stakeholders and to mediate the different or conflicting viewpoints.

Another aspect of that whole situation is that federal agencies tend to have small business enterprise goals, and in order to win a lot of this work, the project team that the firm puts together is made up of a team of firms that includes small business entities as well as large business entities. The CM—who might be a new person right out of college, and not very streetwise as far as the business world goes—needs to recognize that he or she is also facilitating a team of varied business firms that are going to be a part of the success of that project. So again, it is like a multifaceted coordination on the owner's side, and there is also a multifaceted coordination within the project implementation team, and that is another challenge for somebody who is new to the business, to recognize who those folks are and to have the talent to communicate, interact, and coordinate with those various entities, and direct them in a common purpose to get the project done on schedule and on budget.

The U.S. Infrastructure Needs Replacing

LARRY ZIMMERMAN, PE

Business Line Leader for the PM/CM Water Division, Black & Veatch

What is your specialty?

› For 25 years, I owned my own PM/CM companies. I used to own three companies: Construction Dynamics Group, headquartered in Columbia, Maryland; Louis & Zimmerman Value Engineering Company; and CVG International, which was in the U.K. I sold all those companies, went to work, and stayed with the company that bought mine. After three years, I went to Black & Veatch, where I am now the Business Line Leader for the PM/CM Water Division.

What's the future for infrastructure construction?

› Around the country, we are seeing an emphasis on infrastructure renewal for transit, highways, and water, including dams, hydro, tunnels, reservoirs, and wastewater plants. There is a lot of rehab work that is coming about, and these are major, mandated programs. In the water industry, for example, in the Midwest, a lot of utilities, water and sewer parties, are faced with mandated orders to prevent overflows from combined sewers. There are programs all over. Each one of them usually ends up with a PM who is a CM, or else they have another separate CM. It is from cradle to grave, from the initial planning of what to do and how to improve and reduce the overflows of sewers. So you have an eventual impact on the quality of water and an impact on the rivers and streams the sewers discharge in. There is a lot of effort there. Of the whole country, the East Coast has aging cities; their infrastructure is aging. There is a lot of sewer replacement and repair; people go in and replace their whole water systems because they are 100 years old now. You also have water scarcity. You have water in different places, but it is never quite where you want it. Major cities in the West are developing, and they do not have built-in water systems.

What does this mean for your industry and the people who are working in it?

› I think for us, we are seeing an abundance of that type of work. We are also seeing that they are all coming with equipment managers; they want a professional manager to sit with them and manage the programs. That can be for administration of the work, optimizing the cost, or doing it more quickly. The quality aspect is becoming more and more an important element, much more than we have seen in the past.

What are you seeing in innovative financing for these large projects?

› If you look at the banks and investment houses, they are looking for long-term, stable investments. What better long-term, stable investments can you get than, in our case, water? There is a renewed emphasis in investment and a renewed interest in these banks to finance major capital projects. Not so much in the U.S., although there are PPPs all over the place, but definitely overseas these multinational projects are funded by a combination of three or four investment banks. There is definitely a big emphasis there, I think, on the international realm. It is not the lead entity, but not far back is the finance element, that is actually bringing the money and finance to the entire program. So, we are seeing that come into play more and more. For PMs, it is not only delivering the building and designing the capital projects, but it is also managing the finances and social aspects as well. There is more and more emphasis on the social aspects and the mitigation of impacts on constituents. Be it a pipeline, highway, or whatever, the ability to manage that interchange with the public is everywhere. I think the PM has to be a person who can handle many facets. The pure construction, getting the construction done, is a little different, and they just need to be using their skills to deliver.

More PPPs on the Horizon

SHIRLEY YBARRA

Senior Transportation Policy Analyst

REASON FOUNDATION

What's the status of public private partnerships?

› About 23 states have public private partnership statutes. Virginia was one of the first; it is still considered the model bill. I had a major hand in writing it, and it was passed in 1995: The Public Private Transportation Act of 1995.

Almost half of the states have some PPP legislation. All of that says the trend is taking hold; it is an important piece of the puzzle. It is not, and will probably never be, the mainstay. It will always be certain projects that will make sense to be done as a PPP.

Which projects are better suited for PPPs?

› In general, especially when you look at where we have unsolicited possibilities, which is the case in Virginia, they typically need to have some form of revenue generation capability to attract the private sector to invest, whether that be tolls or a new thing—something called availability payments, used frequently in Europe. That is where the agency or owner promises to pay over 25 or 30 years, and they get the money up front from the developer company. Certainly, a CM is going to be looking at whether that makes sense. They need to have some sort of revenue generation capability or guarantee from the owner, something that makes the project work. Remember, these are not low bids. They much more closely resemble a design-build, and certainly most states typically need a design-build as part of a PPP. You need to have that experience to put it together.

What are the risks involved?

› Typically, you will see most of the PPPs have already gone through the environmental process, or virtually gone through the environmental process. That is a risk, for example, that is almost too difficult for the private sector to undertake. Usually the private partners will not take that risk. The CM has to bring the project in on time or under budget. Risk is a real key to these things. Construction risk is often given to the developer. I think as a CM that is probably the big challenge. I would say for a CM they have to be able to be nimble and creative. Sometimes, you dig in the dirt and you find something you didn't expect.

What sort of trends are we seeing in PPPs? Is there anything new or something you foresee coming down in the next five or ten years?

› We saw in two projects in Florida, the Miami Tunnel being the poster child of the use of availability payments. Again, it is the long-term commitment by the owner providing money. It pulls down the amount of up-front commitment, or money, the private company has to put in, because it could not have been done on all private money. It gets the project under way. This is one of the key things I think needs to be said on PPP. It was clearly true on our first PPP, the Pocahontas Parkway. We delivered that project 18 years ahead of what it would have taken us to accumulate the money. Strangely enough, I just read another account of another PPP that did the same thing, 18 years. When you get the federal money in the formula and you have the state transportation money, all that has to be sorted out with formulas. By the time it gets down to a district, you figure out how much money they would get each year, and it would take 18 years before they would ever accumulate that money. You know good and well that money is going to be spent elsewhere; you cannot just put a bank account together and save it for that project. You typically see that the advantage of a PPP is delivering the project sooner. I know time is money; the sooner you can build a project, the less expensive it is.

Do PPPs have any disadvantages?

› The only disadvantage is when the owner goes into a PPP that should never have been a PPP. Otherwise, I don't know of any disadvantages.

5 The Future of Construction Management

THE PRACTICE OF CONSTRUCTION MANAGEMENT is constantly changing to meet the world's demands as well as responding to—and in some cases shaping—new and innovative technology, especially in the area of software development.

Other trends that are changing how CMs do their jobs include innovative financing, public private partnerships, and workforce issues. Another important element is Integrated Program Delivery. Although IPD was touched on in previous chapters, we will discuss it in depth here and explain what it means to those who want to become CMs.

Unlike the traditional design-build project delivery method—whereby the contractor is the leader, taking on the largest share of the responsibility and the blame if something goes amiss—IPD is a Master Builder model whereby an entire team works collectively. Everyone shares in the successes and missteps without the usual finger pointing. This group of collaborators includes the owner, architect, contactor, engineer, CM, fabricators—everyone involved in the project.

As you may have guessed, IPD is not just a different approach to building a project, but a different way of thinking about a project. For some folks, it is such a different way of doing business that they resist the notion straightaway. Others, however, see it as one of the best ways to move construction to the next level.

A Glimpse into the Future

RICHARD FOX

Chairman and Chief Executive Officer

CDM

As far as CMs are concerned, what does the future hold for infrastructure projects?

› I think the general trend is very clear. I think the owners, whether they are public or PPP—which are both owners in some sense—are going to try to do more with less. There is no way to get around the basic drive to do more with less. This means, for instance, that public owners are going to have less and less staff. There is a recent report that shows over the last two years, public owners have reduced their staff by 18 percent with no plans to restore that staff. To me, that bodes well for agency CM to supplement the staff that most of those owners are not going to be involved with.

I also think the PPPs need to keep their costs down, so they will only want staff on board for the period of construction. That means they will turn to private sector CMs instead of staff CMs. I think the trend, whichever owner you are looking at, is the same. I think it is probably true, but I'm not quite as familiar, that the private sector owners have also been curtailing staff. Whether the trend is going to be as clear with independent CMs I don't know, but I feel very strongly it is going to be with public owners and PPP.

I think the CM is going to be asked to participate more heavily in the future on the issue of safety management. Right now, the independent CM or the agency CM has responsibility for means and methods and therefore safety on the site. I do not think in the future, because of the pressure that everyone is going to get on safety, they will be able to occupy that isolated role. I think they are going to be steeped in safety issues. I think their role in safety is going to get much more aggressive. That would include not only injuries to workers and things of that nature, but I think it will get into things like blueprints and certifications.

What about quality issues?

› It will be interesting to see how they sort out the quality issue and how much responsibility they will place on the CM. The CM has the responsibility to QA, but the issue is: When it fails, will they be picking up the responsibility? It will be interesting. I do not think there is a clear trend that the CM is going to get more exposure. If so, you might as well take off the risk of the hard bid. The owners are going to expect more for less; they are going to put pressure on their CMs to keep broadening their scope, even though there will be some resistance to it from a risk perspective.

Do you think CMs are up to the task?

› Yes, I do. I think there are some extraordinarily capable CMs out there that have just been constrained by practice. I know in the safety arena, I feel very comfortable that CMs can step up, provided they do not take on undue risk. I think they are more than capable.

I also think the CMs are going to be brought on much earlier. I'm an advocate that the CM gets brought on at the same time you bring the designers on. That way they can get involved in writing the contract documents and so on about how to build things. They will have more influence, even though they do not select the final means and methods, the sequencing, the contract packaging,

Highway expansion and modernization are a critical national priority. Meeting this need will demand collaboration and CM leadership on complex projects like this restoration of Interstate 64 and Battlefield Boulevard in Hampton Roads, VA. CM by McDonough Bolyard Peck. PHOTO: MCDONOUGH BOLYARD PECK, INC., AND VIRGINIA DEPARTMENT OF TRANSPORTATION (VDOT).

and all of that. The owner will benefit tremendously bringing the CM on earlier to work alongside the designer, almost like an integrated design builder.

This brings to mind IPD. Where do you think we will be with that in the coming years?

› You need to understand one thing: I'm terribly prejudiced on this viewpoint. We believe IPD is the way to go. We're doing that because we are both a designer and a builder; our employees are on both sides of that equation. We are seeing, time in and time out, that IPD is absolutely the way to go. It is the only way you are going to get real productivity out of a construction process that has not gained a lot of productivity over the last few decades.

Are people able to make the emotional leaps needed for IPD to take hold?

› Having lived the experience, internal in the firm, the answer is absolutely yes. What is baffling to me is that once the designers get involved in IPD, it is incredible how creative they can be in terms of thinking about how to build it as well as what they are going to build. I feel very comfortable with design teams. Remember, we have three or four decades where the designer and the builder were at arm's length. There are a lot of emotional issues to get beyond. Once they get beyond them, it is fabulous to watch what happens.

"Smart" transportation systems will use technology to speed traffic flow and, often, to manage variable tolls and volume-based pricing. Shown: A control facility for the New Jersey Turnpike. CM by HNTB, Inc. © ANDY RYAN, PHOTO COURTESY OF HNTB.

What about future technologies and software?

› We are using four-dimensional designs; I think you're going to see five- and six-dimensional designs. The fourth dimension to us is the information around every object placed on the model. The fifth would be time and schedule or money. If you have ever been to Home Depot and done a kitchen, there are two screens: One screen places the counter, floor, or tile on the drawing, and the other screen has the prices. As you build it, like a Lego, you're actually building the prices. That is the fifth key. Then, we are going to truly design the budget. The technology is already there to do it; it is the human processes that we have to continue to work on to exploit the technology. We are labeling it differently than BIM. We are labeling it 4D and 5D to express the extra dimensions. Technology has a critical role. Technology is pretty far down along the path; it is the human work processes we need to continue work on.

PHOTO © TERRY SHAPIRO, COURTESY OF JACOBS.

Integrated Project Delivery

Despite the many advances in construction management, the industry still suffers from the detrimental effects of adversarial relationships, low rates of productivity, high rates of inefficiency and rework, frequent disputes, and lack of innovation. Indeed, the construction industry continues to make great strides in all of these areas—and the situation is improving every day—but the quest for excellence continues as projects still take too long to build and overruns are common.

A relatively new movement to help remedy some of these deficiencies is Integrated Project Delivery, or IPD. The main idea behind IPD is that instead of many contracts and agreements among various players, one overall contract would encompass everyone. The benefit is that all the involved entities would be on the same side from the onset. Instead of maintaining adversarial and often blaming relationships with each other, owners, CMs, trades, and others would be working together from the start, instead of working in their individual silos.

This is, of course, easier said than done, as people tend to cling to their old silo models.

The IPD idea is often credited to a group of Orlando, Florida, businesses that, in the mid-1990s, created the Integrated Project Delivery system. They believed that when they worked as one unit, the owner was better served and the project was completed faster. They reported costs savings as well as less stress than seen in other construction projects. The companies worked together for about five years and actually trademarked the system of IPD in 2005.

IPD is a delivery system that seeks to align interests, objectives, and practices, even in a single business, through a team-based approach. The team's primary members would include the architect and key technical consultants, as well as a general contractor and key subcontractors. The IPD system is a process whereby all disciplines in a construction project work as one firm, creating faster delivery times, lower costs, no litigation, and a more enjoyable process for the entire team—including the owner.

A number of obstacles remain in the path to IPD, though many are being resolved as experience with IPD expands. For example, an IPD contract typically includes a pledge to forgo litigation if things should go awry. Many owners, however, are unwilling to sign away their right to sue their contractor, subcontractors, or CMs. In addition, the American legal system has not traditionally recognized multiparty contracts. These can be very complex: Consider a contract in which Party A promises Party B that Party C will do something by a certain date.

Still, progress is being made. Contracts are being refined, as owners and other stakeholders figure out what works and what doesn't. A growing body of decided court cases is clarifying what types of contracts can succeed. Given the huge benefits to be gained in cost and time savings and trouble-free project execution, all parties have large incentives to work out these complexities. IPD is likely to gain prominence steadily in the years ahead.

PHOTO © TERRY SHAPIRO.

The Future of Integrated Project Management

BLAKE V. PECK, CCM, FCMAA
President/COO
MCDONOUGH BOLYARD PECK, INC.

What's your perspective on IPD?

› IPD grew up in the late 1990s. There was the idea of trying to change things; we had a bunch of alternative delivery methods starting to become popular, in response to some of the issues folks were having with the traditional design/bid/build model. The idea was to see if we could get a collaboration going between the design function and the construction function, like constructability. So, you started seeing the CM at Risk and design/build, where you have early contractor involvement and good input on schedule, cost, and constructability; work things out on paper before they actually go. What we were really doing was integrating the design and construction functions.

What would you like to see, if you could be king of the construction world?

› Conventionally, you define a delivery method by who is actually performing the construction. You can be doing IPD regardless of the delivery method. You could actually be doing it with a general contract, a CM at Risk contract, or a design/build contract. These things are talking about best practices and contracting formats. Do we have a multiparty contract, or do we not? I realize some of this is a bit of semantics, but I worry that if we so narrowly focus and say, "You are not really doing IPD unless you have a multiparty contract," we'll have a sample size of two.

You can perform IPD without being narrowly constrained by a particular set of contract documents or contracting format, and go back to defining delivery methods by who is actually performing the construction, because saying, "You are not doing IPD if you do not have a multiparty contract," seems too narrowly focused. Open the book up to things that we know work. If you are doing CM at Risk, design/build, or a traditional general contract format, these are still best practices and things that work. This is a way of shifting the paradigm.

Instead of having exculpatory language that protects me, if we take an attitude that rather than do what is best for me I do what is best for the project, it is going to work out for everyone in the long run. That is one of the things, at CMAA a few years ago, we got off being the owner's advocate toward being the project advocate. If you do what is right and you make the right decisions for the project, you are going to be taking care of the owner in the long run and bigger picture.

Where do you think we are going to be in five or ten years in CM?

› A lot has been made out of BIM. To me, it is akin to when we first had CAD come in, in the early to mid 1980s. There is that level of expense and getting people used to it, but within a matter of a few years, the technology gets adapted and people get used to working with it. Again, it is a communication device and a way of working out issues. I see technology both in terms of the design function and the project management function. We are again going to get an economy of scale. We are seeing more and more that most of the clients we have are serial builders. They do not do one-off projects, they do multiples, and they have programs.

I think that is why you see the popularity in program management, because people are trying to not learn the same lesson over and over again. I think that is the real thing; technology is going to really expedite that function. Also, like most things with technology, it will change the expectation. Remember twenty years ago if someone didn't get back to you in a day or two, it wasn't a big deal. Now, you send them an email and you're wondering why you haven't heard from them in two hours. That responsiveness has already changed. I see that in delivery times; there is a great need to reduce the entire design construction to operate cycle. I see that being significantly reduced because of the expectations there. The ultimate client does not have the patience to buy into a ten-year or seven-year program; they want to be done in three years. That is what you're seeing.

A good example of this is the federal government, who is the most notoriously slow, patient client ever. Look at their delivery times now, just in say the Base Realignment and Closure program (BRAC), we are delivering hospitals and other buildings in half the time they normally took.

I also see a bigger need for education, besides technological savvy, better credentials, and smarter workforce.

Under the traditional model, with each side in its own camp, the unwanted outcome is practically certain:

- Design effort will be wasted, thereby requiring redesign.
- Construction costs will be higher because general contractors and trade contractors will pad their prices.
- Engineering and safety factors will be extreme.
- Change orders will result.
- Relationships will be adversarial and disputes more frequent.

IPD is not easy to accomplish because it not only goes against the current way of doing business, but it also requires a change in culture and ingrained behavior. These current relationships often are adversarial rather than cooperative, and there are some segments that prefer it that way. It can be argued, for instance, that lawyers make their fees on contentious behavior instead of that which is conciliatory.

All project delivery systems have three basic components: the project organization, the project's "operating system," and the contracts (*commercial terms* in construction parlance) binding the project participants. These are worth looking at in depth.

Rehabilitation of the Tennessee Valley Authority's Blue Ridge Dam in northern Georgia is an example of the large-scale infrastructure modernization that will challenge owners and their CMs in coming decades. PHOTOS COURTESY OF GARNEY CONSTRUCTION COMPANY.

Project Organization

The traditional construction project is organized into three camps, as noted, usually with diverging interests. The only thing they actually agree upon is that they all want the project built. Almost everything else often ends up a matter of confrontation.

With IPD projects, however, these groups share additional traits such as getting involved together during the early stages of design. This changes the project's culture because it stresses the importance of making and keeping commitments, tracking the team's performance, and focusing on improving the reliability of team members' promises.

OPERATING SYSTEMS

The goal is lean operations—getting the most out of every dollar without sacrificing quality, safety, or speed. At the heart of lean operating systems is a change in behavior, especially trust. Trust must be a common thread and is realized through fulfilling commitments. When previously made commitments begin to appear unreliable, participants learn to communicate this earlier rather than later, allowing for more flexibility in the team's response. *A major shift is to engage the team in collaborating to define the problem, rather than critiquing a proposed solution.*

COMMERCIAL TERMS

In most traditional contracts, each participant is acting in its own self-interest, maximizing profits, often to the detriment of the other project players. With IPD, contracts encourage everyone to act as a team, with rewards coming from acting as a team. To accomplish this, IPD projects take a variety of approaches to change the commercial framework of risk allocation and compensation.

For example, during traditional construction projects, certain parties take on risks that they can't control, such as labor dispute delays. In addition, there is no economic benefit for the party not-at-risk for this delay to offer help to the risk-bearing party. With IPD, rather than simply shifting risk (and blame) to each other, members of an IPD team agree to share risk and collectively manage it. They agree to find creative ways of sharing risks and fostering collective risk management.

Three common approaches involve sharing the cost savings or cost overruns against an estimated cost of the work, pooling some portion of the team members' profit and placing it at risk, and/or pooling contingency funds and sharing any amount remaining after project completion. Some people have dubbed this *painsharing:* a common pool of funds used, for instance, to pay for cost overruns. If a project produces an under-run or unexpected savings, these are distributed back to the contributors. *By requiring the major players to share in the risks and benefits resulting from a common fund, this approach gives the key project players a financial interest in helping each other.*

Hill International managed the expansion of the U.S. Capitol Power Plant West Refrigeration Plant. The project team had to remediate 45,000 tons of contaminated soil and work around numerous buried utility, data, and communications lines. PHOTOS COURTESY OF THE ARCHITECT OF THE CAPITOL.

Technology of IPD

Technology plays a large role in making IPD work. Because of its ability to get everyone on the same page, BIM can help foster trust among players because everyone can have the exact same information at the same time. Everyone is working from the same playbook, and nothing is hidden or obfuscated.

Another technology is project management information systems or PMIS. One way to envision PMIS is that it is to the process what BIM is to the product. *A PMIS provides information so the team has a common understanding of the facts: a prerequisite for collaboration.* A PMIS is built around documentation and communication of project-specific information so most of the engine is devoted to that purpose. As the PMIS develops, it will accumulate detailed project information such as cost, schedule, quality, and the team members themselves.

The idea of PMIS is to replace loads of paper with a centralized, comprehensive, near real-time, web-accessible database of electronic project information, available 24/7. Like BIM, it's a resource for the team, created by the team. As with BIM, the main goal is transparency, which bolsters trust and cooperation, which, as we've said, is vital for IPD to work.

Public Private Partnerships

As the need for larger, more expensive, and more complex infrastructure projects grows, and governments are unwilling or unable to fund them outright, public private partnerships, also known as PPP, will become more commonplace.

In simplest terms, a PPP is a business venture—in this case, a construction project—that is funded and built (and often maintained, especially in the case of roads) through a partnership of government and one or more private sector companies. Each party brings its expertise, experience, and financing, and each shares in the risks and rewards. Within these basic parameters, there are many variations. For example, in the case of a highway, the state may offer land or help to obtain it through eminent domain statutes. The private partner may provide financing. When the road is finished, the government has a road that adds to an area's infrastructure, reduces traffic, and helps contribute tax revenue because land adjacent to the road will be developed. Local and state jurisdictions may also gain revenue through payment of police and fire service. The private partner may obtain a return on investment through tolls or by developing adjacent land. Again, this is a simple description of a tool that can be quite complex.

During the 1970s and 1980s, local governments experienced a growth in public debt, and PPPs were thought of as a way for projects to be built with zero public money and little public risk. This pie-in-the-sky model was abandoned very quickly as unrealistic, but the interest in PPPs grew as government and the private sector saw a way to work together for everyone's benefit. Although PPPs are common worldwide, they are very prevalent in Europe and less so—but growing—in the United States.

The main advantage of PPPs is that construction can move along faster, because financing is more easily obtainable, and at less cost to taxpayers. It also maximizes the expertise and resources of both groups.

Disadvantages exist, too. For example, in the case of toll roads, public perception is that the private road owner will be able to raise tolls whenever it wants. This is not true, of course, because there are constraints built into any agreement, but many taxpayers feel they will lose control of public lands and keep paying higher tolls without public oversight. Another disadvantage is that PPP contracts can be very complicated and therefore difficult to enforce because of all the nuances and intricacies. This often can lead to protracted litigation that can slow down projects.

TYPES OF PARTNERSHIPS

Public private partnerships come in a variety of forms, and no two PPP projects are exactly alike. The following comes from *Public-Private Partnerships: Terms Related to Building and Facility Partnerships,* Government Accounting Office, April 1999.

> **O&M: Operations and Maintenance**
>
> A public partner (federal, state, or local government agency or authority) contracts with a private partner to provide and/or maintain a specific service. Under the private operation and maintenance option, the public partner retains ownership and overall management of the public facility or system.
>
> **OMM: Operations, Maintenance & Management**
>
> A public partner (federal, state, or local government agency or authority) contracts with a private partner to operate, maintain, and manage a facility or system providing a service. Under this contract option, the public partner retains ownership of the public facility or system, but the private party may invest its own capital in the facility or system. Any private investment is carefully calculated in relation to its contributions to operational efficiencies and savings over the term of the contract. Generally, the longer the contract term, the greater the opportunity for increased private investment because there is more time available in which to recoup any investment and earn a reasonable return. Many local governments use this contractual partnership to provide wastewater treatment services.
>
> **DB: Design-Build**
>
> A DB is when the private partner provides both design and construction of a project to the public agency. This type of partnership can reduce time, save money, provide

stronger guarantees, and allocate additional project risk to the private sector. It also reduces conflict by having a single entity responsible to the public owner for the design and construction. The public sector partner owns the assets and has the responsibility for the operation and maintenance.

DBM: Design-Build-Maintain

A DBM is similar to a DB except the maintenance of the facility for some period of time becomes the responsibility of the private sector partner. The benefits are similar to the DB with maintenance risk being allocated to the private sector partner and the guarantee expanded to include maintenance. The public sector partner owns and operates the assets.

DBO: Design-Build-Operate

A single contract is awarded for the design, construction, and operation of a capital improvement. Title to the facility remains with the public sector unless the project is a design/build/operate/transfer or design/build/own/operate project. The DBO method of contracting is contrary to the separated and sequential approach ordinarily used in the United States by both the public and private sectors. This method involves one contract for design with an architect or engineer, followed by a different contract with a builder for project construction, followed by the owner's taking over the project and operating it.

A simple design-build approach creates a single point of responsibility for design and construction and can speed project completion by facilitating the overlap of the design and construction phases of the project. On a public project, the operations phase is normally handled by the public sector under a separate operations and maintenance agreement. Combining all three passes into a DBO approach maintains the continuity of private sector involvement and can facilitate private-sector financing of public projects supported by user fees generated during the operations phase.

DBOM: Design-Build-Operate-Maintain

The design-build-operate-maintain (DBOM) model is an integrated partnership that combines the design and construction responsibilities of design-build procurements with operations and maintenance. These project components are procured from the private sector in a single contract with financing secured by the public sector. The public agency maintains ownership and retains a significant level of oversight of the operations through terms defined in the contract.

DBFOM: Design-Build-Finance-Operate-Maintain

With the Design-Build-Finance-Operate-Maintain (DBFOM) approach, the responsibilities for designing, building, financing, operating, and maintaining are bundled together and transferred to private sector partners. There is a great deal of variety in DBFOM arrangements in the United States, and especially the degree to which financial responsibilities are actually transferred to the private sector. One commonality that cuts across all DBFOM projects is that they are either partly or wholly financed by debt leveraging revenue streams dedicated to the project. Direct user fees (tolls) are the most common revenue source. However, others range from lease payments to shadow tolls and vehicle registration fees. Future revenues are leveraged to issue bonds or other debt that provide funds for capital and project development costs. They are also often supplemented by public sector grants in the form of money or contributions in kind, such as right-of-way. In certain cases, private partners may be required to make equity investments as well. Value for money can be attained through life-cycle costing.

DBFOMT: Design-Build-Finance-Operate-Maintain-Transfer

The Design-Build-Finance-Operate-Maintain-Transfer (DBFOMT) partnership model is the same as a DBFOM except that the private sector owns the asset until the end of the contract when the ownership is transferred to the public sector. While common abroad, DBFOMT is not often used in the United States today.

BOT: Build-Operate-Transfer

The private partner builds a facility to the specifications agreed to by the public agency, operates the facility for a specified time period under a contract or franchise agreement with the agency, and then transfers the facility to the agency at the end of the specified period of time. In most cases, the private partner will also provide some, or all, of the financing for the facility, so the length of the contract or franchise must be sufficient to enable the private partner to realize a reasonable return on its investment through user charges.

At the end of the franchise period, the public partner can assume operating responsibility for the facility, contract the operations to the original franchise holder, or award a new contract or franchise to a new private partner. The BTO model is similar to the BOT model except that the transfer to the public owner takes place at the time that construction is completed, rather than at the end of the franchise period.

BOO: Build-Own-Operate

The contractor constructs and operates a facility without transferring ownership to the public sector. Legal title to the facility remains in the private sector, and there is no obligation for the public sector to purchase the facility or take title. A BOO transaction may qualify for tax-exempt status as a service contract if all Internal Revenue Code requirements are satisfied.

BBO: Buy-Build-Operate

A BBO is a form of asset sale that includes a rehabilitation or expansion of an existing facility. The government sells the asset to the private sector entity, which then makes the improvements necessary to operate the facility in a profitable manner.

Developer Finance

The private party finances the construction or expansion of a public facility in exchange for the right to build residential housing, commercial stores, and/or industrial facilities at the site. The private developer contributes capital and may operate the facility under the oversight of the government. The developer gains the right to use the facility and may receive future income from user fees.

While developers may in rare cases build a facility, more typically they are charged a fee or required to purchase capacity in an existing facility. This payment is used to expand or upgrade the facility. Developer financing arrangements are often called capacity credits, impact fees, or extractions. Developer financing may be voluntary or involuntary depending on the specific local circumstances.

EUL: Enhanced Use Leasing or Underutilized Asset

An EUL is an asset management program in the Department of Veterans Affairs (VA) that can include a variety of different leasing arrangements (e.g., lease/develop/operate, build/develop/operate). EULs enable the VA to long-term lease VA-controlled property to the private sector or other public entities for non-VA uses in return for receiving fair consideration (monetary or in-kind) that enhances VA's mission or programs.

LDO or BDO: Lease-Develop-Operate or Build-Develop-Operate

Under these partnership arrangements, the private party leases or buys an existing facility from a public agency; invests its own capital to renovate, modernize, and/or expand the facility; and then operates it under a contract with the public agency. A number of different types of municipal transit facilities have been leased and developed under LDO and BDO arrangements.

Lease/Purchase

A lease/purchase is an installment-purchase contract. Under this model, the private sector finances and builds a new facility, which it then leases to a public agency. The public agency makes scheduled lease payments to the private party. The public agency accrues equity in the facility with each payment. At the end of the lease term, the public agency owns the facility or purchases it at the cost of any remaining unpaid balance in the lease.

Under this arrangement, the facility may be operated by either the public agency or the private developer during the term of the lease. Lease/purchase arrangements have been used by the General Services Administration for building federal office buildings and by a number of states to build prisons and other correctional facilities.

Sale/Leaseback

This is a financial arrangement in which the owner of a facility sells it to another entity, and subsequently leases it back from the new owner. Both public and private entities may enter into sale/leaseback arrangements for a variety of reasons. An innovative application of the sale/leaseback technique is the sale of a public facility to a public or private holding company for the purposes of limiting governmental liability under certain statues. Under this arrangement, the government that sold the facility leases it back and continues to operate it.

Tax-Exempt Lease

A public partner finances capital assets or facilities by borrowing funds from a private investor or financial institution. The private partner generally acquires title to the asset, but then transfers it to the public partner either at the beginning or end of the lease term. The portion of the lease payment used to pay interest on the capital investment is tax exempt under state and federal laws. Tax-exempt leases have been used to finance a wide variety of capital assets, ranging from computers to telecommunication systems and municipal vehicle fleets.

Turnkey

A public agency contracts with a private investor/vendor to design and build a complete facility in accordance with specified performance standards and criteria agreed to between the agency and the vendor. The private developer commits to build the facility for a fixed price and absorbs the construction risk of meeting that price commitment. Generally, in a turnkey transaction, the private partners use fast-track construction techniques (such as design-build) and are not bound by traditional public sector procurement regulations. This combination often enables the private

> partner to complete the facility in significantly less time and for less cost than could be accomplished under traditional construction techniques.
>
> In a turnkey transaction, financing and ownership of the facility can rest with either the public or private partner. For example, the public agency might provide the financing, with the attendant costs and risks. Alternatively, the private party might provide the financing capital, generally in exchange for a long-term contract to operate the facility.

This summary illustrates the diversity (and potential complexity) of the financing and construction delivery systems being explored by innovative owners today. The future will put a premium on this type of creativity, and further strengthen the role of a qualified, owner-focused CM in bringing projects into reality.

Century City in Los Angeles included some of the first skyscrapers to be built in the city after revision of earthquake-related restrictions in the city building code. Tishman Construction served as CM, incorporating new building materials and methods. PHOTO COURTESY OF TISHMAN CONSTRUCTION.

Innovative Financing

As the worldwide need for construction grows, governments and the private sector both see the need to go beyond traditional financing methods. One way, of course, is PPP, but there are others. First, a definition is in order: The World Bank describes innovative financing (sometimes called alternative financing) this way:

> Innovative financing involves non-traditional applications of solidarity, PPPs, and catalytic mechanisms that (i) support fundraising by tapping new sources and engaging investors beyond the financial dimension of transactions, as partners and stakeholders in development; or (ii) deliver financial solutions to development problems on the ground.

As you can see, innovative financing can combine many different ideas such as IPD and public private partnerships, with the goal of getting projects off the ground quickly and where they are needed. Although the World Bank's main interest is in helping developing countries, these innovative financing methods can be used anywhere in the world.

DESIGN-BUILD-OPERATE-MAINTAIN

In traditional construction projects, the CM often acts in two capacities: design and build. In a popular newer model, CMs design, build, operate, and maintain. Although most often used for road and rail construction, DBOM can also be seen in building projects. CMs may be in the best position to operate and maintain a project because of their extensive knowledge of how it was financed and built.

According to the U.S. Department of Transportation: "The design-build-operate-maintain model is an integrated partnership that combines the design and construction responsibilities of design-build procurements with operations and maintenance. These project components are procured from the private sector in a single contract with financing secured by the public sector. This project delivery approach is practiced by several governments around the world and is known by a number of different names, including 'turnkey' procurement and build-operate-transfer (BOT)."

As with all financing methods, there are advantages and disadvantages. The main advantage of the DBOM approach is that it combines responsibility for usually disparate functions—design, construction, and maintenance—under a single entity. This allows the private partners to take advantage of a number of efficiencies, including the ability to tailor the project design to the construction equipment and materials that will be used.

Because it will maintain the project, the DBOM team must establish up front a long-term maintenance program and estimate its cost. This is not something done in the design-build model by the designer and builder, who, not so surprisingly, are in the best position to do so. The team's detailed knowledge of the project design and the materials allows it to develop a tailored maintenance plan that anticipates and addresses needs as they occur, thereby reducing the risk that issues will go unnoticed or unattended and then deteriorate into much more costly problems.

This focus on the true "life-cycle cost" of a project, rather than simply the initial cost of construction, goes hand in hand with today's pervasive attention to sustainability. Long-term operating costs are heavily influenced by the cost of heating or cooling a building; by its water consumption and the volume of waste water it generates; by its maintenance needs and ongoing supply requirements, and so on. When dealing with the responsibility of operating a facility over the long term, the project team has a strong incentive to explore options such as capturing rainwater, reusing wastewater, managing sunlight exposures, and minimizing auto traffic on a site.

DBOM does present some dangers for owners, however. Owners who are not used to this approach must make certain that they specify exactly all of the standards of design, build, operate, and maintain that they want. And it must be done before the project begins. This can be difficult because some projects, such as roads and rails, can have contracts extending up to 20 years or more. Unless the owners are very careful, they may not receive the operation and maintenance service that they desire because these may not be precisely articulated. Once they have signed the contract, they have given up their control, perhaps for decades.

OTHER FINANCING METHODS

Two other forms of innovative financing are worth mentioning: sale-and-leaseback and user fees.

Sale-and-leaseback, sometimes shortened to leaseback, is an arrangement whereby one entity sells an asset and leases it back for a long term. In essence, the seller receives money for the asset, no longer owns it, but has the ability to use it. This is mainly done for fixed assets such as land. Leasebacks can be very useful when the landowner needs fast money to develop the property. In some arrangements, after the land has been developed, the original owner can buy back the land with money that has been obtained through a revenue-generating building.

User fees are what the name implies. The main users of the facilities, often roads and bridges, pay to use them. This has become commonplace as a way to retroactively fund the building of toll roads and parks. The main advantage is that it puts the focus on those who actually use a facility, which many people find ultimately equitable. The downside of user fees is that if they are not priced appropriately, or calculations were wrong, there may not be enough fees to cover the construction and/or operating expenses. In this case, outsiders—often taxpayers, in the case of a road or bridge—may have to pitch in.

Technology

As in most industries, technology will shape the future as the future shapes technology. Construction management is no different. In general, CM technology takes the form of new construction techniques and materials and in software development.

Public infrastructure projects large and small, illustrating the range and diversity of work awaiting professional construction managers. All projects had CM by ARCADIS. From the top: The Port of Long Beach, one of the world's busiest, has been expanded several times, including a new Pier C complex; Yuma County Library, AZ; Santan Vista Water Treatment Plant, Gilbert, AZ, which serves Gilbert and Chandler, AZ; and San Elijo Lagoon Nature Center, CA. FROM TOP, PHOTOS COPYRIGHT BY: HELIPHOTO; ARCADIS U.S., INC.; ARCADIS U.S., INC.; PABLO MASON.

Looking Ahead in Technology

ERIC LAW

Founder and President

EADOC SOFTWARE

What technology trends are you seeing?

› The trend of fewer firms with an increase of knowledge workers will continue. There will be continued acquisition and consolidation as the industry goes through tough economic times.

Agencies are looking for CM firms to bring value to their projects in terms of engineers with project management expertise who can execute a project. They are not looking to pay for administrators to push paper.

Technology adoption will continue to increase dramatically as firms look for ways to increase their competitive advantages and cut costs.

What technologies or tools will really "come into their own" in the coming years?

› True, collaborative web-based applications delivered over the Internet giving members of project teams access to all their project information 24/7 from anywhere in the world. Companies can no longer afford to pay staff to search through filling cabinets, log in to complex VPNs, or manually re-enter data into legacy enterprise software applications. The days of every project participant using their own PM solution for information management are shrinking. As teams become more integrated and projects increase in complexity, it becomes paramount for the team to be using a single collaborative application for exchanging and controlling the flow of information.

Who may be left behind?

› There will always be firms averse to technology and improving productivity. The market for these firms will continue to shrink exponentially.

Are there big gaps that will be filled in terms of users' and CMs' needs?

› You are going to see continued integration and collaboration between vendor applications. This will progress slowly until large numbers of users migrate from old legacy enterprise applications that have no integration capability. Also, standardization is still very weak in the AEC industry. I would expect to see this area start to really take off in 2012/2013.

What's the perfect world for you?

› Government agencies would lay out long-term (20–40 year) infrastructure plans with solid funding, regardless of current political winds and special interest groups. They would implement projects bringing the greatest value to the area served using the most efficient methods to design and construct the projects.

For example, CMs must be aware of innovations taking place in construction materials such as the growing interest in the substitution of steel for wood in many structures. Likewise, many construction projects are employing new kinds of foundations. For example, we're seeing increased use of soil mitigation instead of traditional pilings. In areas where the ground may be soft because of sand, for instance, buildings are often built on pilings that are hammered into the ground to great depths because the soil can't support a building or bridge. Instead of the pilings, some CMs are opting for soil mitigation, which digs out an area and replaces it with rock or some mixture of sand and rock. This allows a structure to be built on solid ground. The advantages are many, including lower cost, but also environmental benefits. Instead of filling the ground with concrete and steel, natural rock is used. In addition, the carbon footprint is lowered because heavier, pile-driving equipment is not used.

The Westin Hotel in New York harmonized with the "entertainment" flavor of nearby Times Square. PHOTO COURTESY OF TISHMAN CONSTRUCTION.

"If You Can't Measure It, You Can't Manage It"

JOSH KANNER

Vice President of Marketing and Business Development

Vela Systems

What does Vela do?

› It's all about bringing documents and work flow out to the point of construction. It's all about bringing technology to where the work is done out in the field. If you look at some research from the Construction Industry Institute, about 25 cents of every dollar is wasted out in the field. If you look at where the construction technology dollars are spent, 90 percent of the dollars are spent for stuff either in the trailer or the back office. The technology is not getting out to where the waste is and where the dollars are spent, which is out on the job site. Vela Systems is all about how you can actually bring technology out there with you when you go out into the field and do the work.

Specifically, it is a combination of three things. One, both document synchronization and control available out in the field whether you're online connected to the Internet or not. You can bring up all of your construction documents, owners can bring up manuals, and all the information you would need out in the field.

Part two is a whole set of management capabilities as well as specific work flows for checklists. Checklists are used to do things like run your quality control programs, run safety programs, and even do league inspections. All the work that goes out in the field can be done through checklists. Vela Systems in the second component of what we do, is a whole set of checklists for the field, which can then be edited or tailored however you want for a particular project. Then, perhaps even more importantly, it is a whole infrastructure for a company to manage how they do their safety and quality programs using checklists and documents.

Item number three is something you can think of as issues and issue management. These are things you might find when you're walking the job site, whether it is a work-to-complete issue, RFI, another question, or anything you might find out on the job site that you can record either with a photo, verbal description, or voice record it. Again, whether you are connected to the Internet or not, you can have that all go into one master issues list, where it can be managed by whoever needs to deal with it and whatever they need to do. Those are the three workflow components.

I think the bigger-picture value is really all about managing the field from both different levels. We also have, which is very important from a CM standpoint, a whole series of dashboards that measure the effectiveness of your programs for things like quality control, safety, and even using BIM in the field. All that rolls up to a dashboard with metrics, which allows you to benchmark your own projects, teams, subs against each other. It is a cloud-computing-based software.

A very exciting thing for us is that some leading construction managers are now starting to use our data to benchmark, not only within their companies, but across companies, including things like quality control, conformance rates, frequency of inspections, safety, and other indicators of quality.

Safety is a big one that people like to benchmark out in the field. If I'm a CM out in the field, how can I address that? Can I use my smart phone perhaps?

› There are two modes of accessing this data. One is if you have Internet connection, so you could use your smart phone through the Internet, or you could use a laptop computer or tablet PC, anything that has Internet access. The other mode is without Internet access. Oftentimes, a lot of our field users, such as field supervisors, PMs, project engineers, etc., use Vela in a disconnected mode. The disconnected mode for Vela is on the Apple iPad.

Why did you believe there was a need for this?

› The company started in 2005 because of the frustrations of a CM who had spent 10 years out in the field putting out half-sized sets of drawings and forgetting the right one he needed or writing down a set of things that needed to happen and then having to wait for them to be typed up. He literally went one day to return his rental car, and the rental car return person scanned his car with a device. The CM, Adam Lamansky, who is the other founder of Vela, thought, "This is ridiculous. I'm on a project now where the courier costs are in the tens of thousands of dollars a day or more to the owner. The rental car guy has something to use in the field, and every time I go out to the field I have nothing. I have to come back and waste two to three hours a day just trying to communicate what I've seen on the job site to everybody who needs to know about it." There is a lot of waste there. That is when Vela got started. The whole concept was developed farther at a joint-degree program here in Boston with the Harvard Graduate School of Design and the MIT Center for Real Estate. We wound up building the company off the core concept that there are a lot of ways to be more efficient in the field. That was where we started; it was really in the field. Interestingly, over the last five years, the value of what we are doing has really evolved from person productivity and product acceleration; those were the two keys values for Vela and still are.

"If you are not measuring it, you can't manage it" is the classic saying. That has been a very powerful aspect of using our software. Because the information is being generated, whether it is checklists, issue counts, or issue closures, in the field where the work gets done, there is no filter for the information. It is not filtered by someone's mind or filtered by what someone typed in. As a result, we are able to create new sources of information around quality, safety, and around the performance of teams and trades—sources that never existed before. It is a very exciting time, from our perspective, for CM because you can do some real measurement, and thus real management of your construction.

The faster-acting changes are being seen in software and communications development. By most measures, BIM, as we've seen in previous chapters, is the most exciting software development to enter the construction world, and it will continue to evolve, grow, and gain acceptance. However, getting BIM into the field so CMs can use it wherever they are remains a challenge, especially because people use diverse platforms, including smart phones.

As more and more players join the CM's team, and collaboration increases, reaching everyone with data and graphics continues to be a challenge. So, as software becomes more sophisticated, telecommunications will have to keep pace.

Thinking Like Owners

JON ANTEVY

CEO and Founder

e-Builder

Tell me about e-Builder.

› We provide an online project management system. For example, a construction manager can log in to e-Builder and access all of the project costs that have been incurred to date, compare that to the budget, and see how their project is tracking. If they care about the design aspects as they are going through design review, they can access all of the drawings, plans, and specifications that the engineers or architects are posting in the system. They can comment and collaborate back and forth on those. Also, they can spot potential issues when they are small, before they become big issues.

Suppose I'm a CM, I'm working with an owner, and we already have BIM software that we are using. Do you take that and put it online?

› Not really; we work a lot and integrate with BIM software. A typical BIM software that an architect or engineer would use is Rivet. They are developing their designs in Rivet BIM, and they need to communicate that to multiple stakeholders. The files are very large, so we provide a place that gives unlimited storage. We do not care how big the files are. Then, anytime changes are made to those files, it is automatically communicated to those who need to know about it. It also provides an audit trail of who did what to this file. It always presents the most current, up-to-date file to the whole team.

Do you restrict who can make changes? Is that part of your service?

› Yes, there is quite a detailed list of permissions within the system. We always like to remind people that if they do not have access to something, they will not even see it on their system.

Do you also have features whereby, for example, a user would have access to look at something, but not to make changes?

› Correct, there is read-only.

As far as my device goes, can I look at it on a PC, Mac, phone, tablet, whatever. . . ?

› Yes to all of those. The most popular are the droids, iPads, and iPhones. It is 100 percent browser compliant. I am a big iPhone guy, I have an iPad at home, and I often access the system through there.

Can you give me an example of where this would really save money for an owner?

› There is one that comes to mind. It is a hundred-million-dollar school bond program over five years. Year one, the CM got the system; he got e-Builder and established processes for everything: The way that changes to the contract are approved. Scott had a specific process for that. The way that invoices are approved, received, paid, Scott had a process for that. The reports that are important so you have visibility of what is going right versus wrong . . . he had a process for that. That was all baked into e-Builder.

Two years into the program, the CM approached the owner and said: "I have to leave." He went to work for his dad's contracting business to help him out. The owner was in a dilemma. He looked for someone of the CM's caliber, but could not find it. Then he said to himself, "Hold on a second, I needed the horsepower and the CM's brainpower in the beginning, but now that all of those processes have been baked into e-Builder, I don't need a high-powered person like that." So, he asked the program manager firm to give him someone less expensive, frankly, which they did. They then paid another $1,000 a month for an intern from a university who is now the e-Builder administrator. With three years left on the program, he issued a $367,000 deductive change order. He just presented this at our user conference, which is why it is fresh in my mind.

Where do you see things going as far as technology and tools are concerned for CMs?

❯ I definitely see the owners becoming smarter with these tools, especially the large owners who are involved in building programs where they are effectively repeat builders. That includes government, K–12, higher education, healthcare. CMs that are working for these owners really need to understand how to work with them. We are seeing a lot of PPP, public private partnership, which is really becoming interesting. Now that there is less money that the government has, they are looking towards the private market. It is effectively turning a CM into an owner, or even a contractor into an owner. For that reason, these contractors and CMs have to start thinking like owners as well.

Workforce Issues

Like every other profession, construction management has its share of workforce issues. The most important among them is a lack of trained and experienced CMs to fill the growing void caused by increasing construction projects as well as the need to replace those people who are retiring. There are not enough schools turning out CMs, although the number of schools teaching construction management is growing. This is not just a U.S. problem, but a worldwide phenomenon.

According to Jesus M. de la Garza, PhD, at Virginia Tech's Myers-Lawson School of Construction:

> The future demand for the Construction Management profession is on the upswing; however, to ensure a steady supply of CM professionals, universities need to develop stronger and more rigorous curriculums to keep up with the ever expanding expectations. Advanced graduate degrees in Construction Engineering and Management, Construction Management, and/or Project Management, for example, should become the expected credentials of those practicing the profession. These advanced degrees, typically a Master's Degree,

> combined with certifications will, without a doubt, elevate the standing and reputation of the profession. The Master's degrees I refer to are analogous to the successful Masters in Business Administration (MBA). The principle is simple: The more education we require from those practicing CM, the higher the standing of and respect for the CM profession.

Exacerbating the issue is that many students are not encouraged at the high school level to consider a career in construction management. This is often because guidance counselors are not familiar with the profession. They often equate it with construction trades, which, traditionally, they have recommended to non-college-bound students.

These shortages are not unique to construction; in fact, nearly every major American industry faces the same situation.

The widely publicized need for massive new investment in America's infrastructure—roads, bridges, rail, transit, waterways, electrical grid, schools, and other needs—is also expected to accelerate the demand for construction professionals and CMs.

PHOTO BY JIM WINN, PHOTO@JIMWINN.COM.

Tackling the Workforce Shortage

PAMELA R. MULLENDER

President/CEO

ACE Mentor Program of America

Describe the ACE Mentor Program and the workforce challenges you hope to meet with it.

› The ACE Mentor Program, Architecture Construction and Engineering Program, was started as an idea in 1991. It started as a result of someone's concern that America was not producing the number of engineers anymore that we used to. So, it actually started because of what they saw as a pending workforce issue for the integrated construction industry. As it was formed, the founders decided that instead of just doing engineering they would make it a program that could reach out to a variety of interests, a variety of kids with many different interests, and teach them about the integrated construction industry through a mentoring program where the students had to mimic what the mentors did in real life.

ACE is an after-school program which meets with students about 15 to 18 times during the course of the school year for at least two hours every time. We have a team of mentors made up from individuals from architecture, construction, different types of engineering, and some subcontractors and skilled trade workers are involved. We match that mentor team with a team of students, and they work together all during the school year. By the end of the time they are with us, they must create, all the way from concept all the way through to delivering it to a client, a project of their own. In doing so, they learn a little bit about each of the different facets of the construction industry.

What are the workforce challenges facing us?

› In 2007, we estimated that the industry would be over a million people short by 2012. That workforce issue has not gone away, even though the construction industry as a whole has faced many layoffs.

If you look at the industry, you see that it is very fragmented. Parents and guidance counselors just do not know that much about the industry. We, as an industry, have not done a good job letting people know it is available to them. Because of our outreach, we can give kids a better perception of what is going on in the construction industry. That is not necessarily the fault of the guidance counselor, when I say that. I was a schoolteacher many years ago. I would not want to be a schoolteacher or a guidance counselor today, when you look at some of the schools in New York where they have one guidance counselor for 1,500 students. There are only so many times you see a student during the course of the year, and there is only so much you can tell them. Doctors and lawyers are always very appealing to family members, when you can say, "My son is going into the legal field." Not "My son is going to be a CM."

Is this a global phenomenon? Or is it just in the U.S. we are seeing this shortfall?

› That answer is two parts. I know more about what is going on in the U.S. I do not know what is going on, as much, in say China or India. Recently, I read an article in the *New York Times* about the terrible infrastructure in India. The roads are deteriorating; they are worse than our highway system. The reason is because they do not have any civil engineers over there. However, both China and India every year produce more graduates in engineering than the U.S. China produces 300,000; India 200,000; and we graduate 75,000. The answer to the other part of that question is: Most of the companies we are working with are global. When they go overseas, they also have issues with finding workforce.

Where do you think we will be in five or ten years?

› I think in five years, let's say by 2015, we will still have a shortage of construction-related workers. That is everything from pipe fitters right up through to the world-renowned architects and engineers. I believe we have only just begun to see the problem. I think it is kind of veiled right now because of the economic situation. When money starts freeing up from the banking side of things to fund these projects, we will be hurting for employees.

PHOTO BY RICH SARLES, COURTESY OF KEVILLE ENTERPRISES, INC.

APPENDIX

Code of Ethics

Like other professionals, construction management leaders have established a code of ethics. The following comes from CMAA:

As a professional engaged in the business of providing construction and program management services, and as a member of CMAA, I agree to conduct myself and my business in accordance with the following:

1. **Client Service.** I will serve my clients with honesty, integrity, candor, and objectivity. I will provide my services with competence, using reasonable care, skill and diligence consistent with the interests of my client and the applicable standard of care.
2. **Representation of Qualifications and Availability.** I will only accept assignments for which I am qualified by my education, training, professional experience and technical competence, and I will assign staff to projects in accordance with their qualifications and commensurate with the services to be provided, and I will only make representations concerning my qualifications and availability which are truthful and accurate.
3. **Standards of Practice.** I will furnish my services in a manner consistent with the established and accepted standards of the profession and with the laws and regulations which govern its practice.
4. **Fair Competition.** I will represent my project experience accurately to my prospective clients and offer services and staff that I am capable of delivering. I will develop my professional reputation on the basis of my direct experience and service provided, and I will only engage in fair competition for assignments.
5. **Conflicts of Interest.** I will endeavor to avoid conflicts of interest; and will disclose conflicts which in my opinion may impair my objectivity or integrity.
6. **Fair Compensation.** I will negotiate fairly and openly with my clients in establishing a basis for compensation, and I will charge fees and expenses that are reasonable and commensurate with the services to be provided and the responsibilities and risks to be assumed.
7. **Release of Information.** I will only make statements that are truthful, and I will keep information and records confidential when appropriate and protect the proprietary interests of my clients and professional colleagues.
8. **Public Welfare.** I will not discriminate in the performance of my Services on the basis of race, religion, national origin, age, disability, or sexual orientation. I will not knowingly violate any law, statute, or regulation in the performance of my professional services.

9. **Professional Development.** I will continue to develop my professional knowledge and competency as Construction Manager, and I will contribute to the advancement of the construction and program management practice as a profession by fostering research and education and through the encouragement of fellow practitioners.

10. **Integrity of the Profession.** I will avoid actions which promote my own self-interest at the expense of the profession, and I will uphold the standards of the construction management profession with honor and dignity.

Glossary

Addendum

A supplement to documents, issued prior to taking receipt of bids, for the purpose of clarifying, correcting, or otherwise changing bid documents previously issued.

Additional Services

Services provided in addition to those specifically designated as basic services in the agreement between the owner and CM. Also known as supplemental services.

Agency

A legal relationship by which one party is empowered and obligated to act on behalf of another party.

Agency Construction Management

A form of construction management performed in a defined relationship between the CM and owner. The agency form of construction management establishes a specific role of the CM acting as the owner's principal agent in connection with the project/program.

Agreement

A document setting forth the relationships and obligations between two parties, as the CM and owner or contractor and owner. It may incorporate other documents by reference.

Apparent Low Bidder

The bidder who has submitted the lowest bid for a division of work described in bid documents, a proposal form, or proposed contract.

Approved Bidders List

The list of contractors that have been prequalified for the purpose of submitting responsible, competitive bids.

Approved Changes

Changes in the contract documents that have been subjected to an agreed-upon change approval process and have been approved by the party empowered to approve such changes. See "Change Order."

As-Built Drawings

Drawings (plans) that show the work, as actually installed. Also known as record drawings.

At-Risk Construction Management

A delivery method that entails a commitment by the construction manager to deliver the project within a guaranteed maximum price (GMP). The construction manager acts as consultant to the owner in the development and design phases, but as the equivalent of a general contractor during the construction phase. When a construction manager is bound to a GMP, the most fundamental character of the relationship is changed. In addition to acting in the owner's interest, the construction manager also protects him/herself.

Basic Services

Scope of service as defined in the original agreement between the owner and CM as basic services.

Beneficial Occupancy

The use of the constructed facility by the owner prior to final completion of the construction.

Bid

An offer to perform the work described in contract documents at a specified cost.

Bid Bond

A pledge from a surety to pay the bond amount to the owner in the event the bidder defaults on its commitment to enter into a contract to perform the work described in the bid documents for the bid price.

Bid Documents

The documents issued to the contractor(s) by the owner that describe the proposed work and contract terms. Bid documents typically include drawings, specifications, contract forms, general and supplementary general conditions, proposal or bid forms, and other information.

Biddability

The degree to which a set of bid documents could be reasonably expected to permit a bidder to establish a competitive price to perform the work as defined in the bid documents.

Biddability Review

A formal review of the contract documents, addenda, and reference documents to be accomplished with respect to the local construction marketplace and the bid packaging strategy so as to eliminate ambiguities, errors, omissions, and contradictions, for the purpose of minimizing bid prices in the procurement phase and disputes during construction.

Bond

A pledge from a surety guaranteeing the performance of the obligation defined in the bond, including the completion of work or payment of the bond amount to the obligee (owner or contractor) in the event of a default, or nonpayment by a principal (contractor or subcontractor), as with bid, performance and labor, and material bonds.

Bonus

Additional compensation paid or to be paid to the contractor by the owner as a reward for accomplishing predetermined objectives that are over and above the basic requirements of the contract between the owner and contractor.

Budget

The dollar amount allocated by the owner for a project/program.

Budget Estimate

An estimate of the cost of work based on preliminary information, with a qualified degree of accuracy.

Changed Conditions

Conditions or circumstances, physical or otherwise, that differ from the conditions or circumstances on which the contract documents were based.

Change Order

A written agreement or directive between contracted parties that represents an addition, deletion, or revision to the contract documents; identifies the change in price and time; and describes the nature (scope) of the work involved. Also known as a contract modification.

Claim

A formal demand for compensation, filed by a contractor or the owner with the other party, in accordance with provisions of the contract documents.

CM Fee

A form of contractual payment for services, whereby the CM is paid a fee for services performed.

Code of Accounts

The owner's written description of the cost elements of the project, used for the owner's accounting purposes.

Commissioning

Start up, calibration, and certification of a facility.

Constructability

The ease with which a project can be built, based upon the clarity, consistency, and completeness of the contract documents for bidding, administration, and interpretation to achieve overall project objectives.

Constructability Reviews

The process of evaluating the construction documents for clarity, consistency, completeness, and ease of construction to facilitate the achievement of overall project objectives.

Construction Budget

The sum established, normally during the pre-design or design phase, as available for construction of the project.

Construction Contract Documents

The documents that provide the basis for the contract entered into between parties. They typically include the bid documents updated to reflect the agreement between the owner and the contractor(s).

Construction Cost

See "Cost of Construction."

Construction Management

A professional management practice applied to construction projects, from project inception to completion, for the purpose of controlling time, cost, scope, and quality.

Construction Management Plan

The written document prepared by the construction manager (CM), that clearly identifies the roles, responsibilities, and authority of the project team and the procedures to be followed during construction.

Construction Manager (CM)

An organization or individual with the expertise and resources to provide construction management services.

Construction Schedule

A graphic, tabular, or narrative representation or depiction of the time of construction of the project, showing activities and duration of activities in sequential order.

Contingency

An amount of money reserved by the owner to pay for unforeseen changes in the work or increases in cost.

Contract Administration

The function of implementing the terms and conditions of a contract, based upon established systems, policies, and procedures.

Contractor

The organization or individual who undertakes responsibility for the performance of the work, in accordance with plans, specifications, and contract documents, providing and controlling the labor, material, and equipment to accomplish the work.

Cost Control

The function of limiting the cost of the construction project to the established budget based upon owner-approved procedures and authority.

Cost Management

The act of managing all or partial costs of a planning, design, and construction process to remain within the budget.

Cost of Construction

All costs attributed to the construction of the project, including the cost of contracts with the contractor(s), construction support items, general condition items, all purchased labor, material, and fixed equipment.

Critical Date Schedule

See "Milestone Schedule."

Critical Path Method (CPM)

A scheduling technique used to plan and control a project. CPM combines all relevant information into a single plan defining the sequence and duration of operations, and depicting the interrelationship of the work elements required to complete the project. The critical path is defined as the longest sequence of activities in a network that establishes the minimum length of time for accomplishment of the end event of the project. Arrow Diagramming Method (ADM) and Precedence Diagramming Method (PDM) are both common forms of CPM scheduling.

Design-Build

A project delivery method which combines architectural and engineering design services with construction performance under one contract agreement.

Designer

The individual or organization that performs the design and prepares plans and specifications for the work to be performed. The designer can be an architect, an engineer, or an organization which combines design services with other professional services.

Design—Final Stage

The stage of the design process when drawings and specifications are completed for construction bid purposes. It is preceded by the preliminary design stage, and followed by the procurement phase. This designation is used by designers for the last part of the design process prior to procurement.

Design—Preliminary Stage

The transition from the schematic stage to the completion of design development. During this stage, ancillary space is developed and dimensions are finalized. Outline specifications are developed into technical specifications; sections are delineated, and elevations are defined. Also known as design development.

Design—Schematic Stage

Traditionally the first stage of the designer's basic services. In the schematic stage, the designer ascertains the requirements of the project and prepares schematic design studies consisting of drawings and other documents illustrating the scale and relationships of the project.

Direct Costs

The field costs directly attributed to the construction of a project, including labor, material, equipment, subcontracts, and their associated costs.

Drawings

Graphic representations showing the relationships, geometry, and dimensions of the elements of the work.

Estimated Cost to Complete

The current estimate of the remaining costs to be incurred on a project at a specific point in time.

Estimated Final Cost

The anticipated cost of a project or project element when it is complete. The sum of the cost to date and the estimated cost to complete.

Fast Track

The process of dividing the design of a project into subphases in such a manner as to permit construction to start before the entire design phase is complete. The overlapping of the construction phase with the design phase.

Field Order

An order issued at the site by the owner or CM to clarify requirements and/or require the contractor(s) to perform work not included in the contract documents. A field order normally represents a minor change, not involving a change in contract price or time, and may or may not be the basis of a change order.

Final Completion

The date on which all the terms of the construction contract have been satisfied.

Float

Contingency time that exists on a scheduled activity. It represents the amount of time that activity may be delayed without affecting the end date of the schedule. It is measured by comparing the early start and late start, or early finish and late finish, dates of an activity.

Force Account

Directed work accomplished by the contractor outside of the contract agreement, usually paid for on a time and material basis.

General Conditions

A section of general clauses in the contract specifications that establish how the project is to be administered. Included are obligations such as providing temporary work, insurance, field offices, etc.

Guarantee

A legally enforceable assurance by the contractor and/or a third party of satisfactory performance of products or workmanship during a specific period of time stated and included in the contract.

Guaranteed Maximum Price

A contractual form of agreement wherein a maximum price for the work is established based upon an agreed-to scope.

Lien

A claim, encumbrance, or charge against or an interest in property to secure payment of a debt or performance of an obligation.

Life-Cycle Cost

All costs incident to the planning, design, construction, operation, maintenance, and demolition of a facility or system, for a given life expectancy, all in terms of present value.

Liquidated Damages

An amount of money usually set on a per day basis, which the contractor agrees to pay the owner for delay in completing the work, in accordance with the contract documents.

Long Lead Item

Material or equipment having an extended delivery time. Such items may be considered for early procurement and purchase under separate contract, to facilitate on-time completion of the project.

Long Lead Time

The extended time interval between purchase and delivery of long lead items.

Low Bidder

The responsible bidder who has submitted the lowest bid that is determined to be responsive to the request for bids for a division of work described in a bid document, proposal form, or contract.

Lump-Sum Fee

A fixed amount that includes the cost of overhead and profit paid, in addition to all other direct and indirect costs of performing work.

Master Schedule

An executive-level summary schedule identifying the major components of a project, their sequence, and their durations. The schedule can be in the form of a network, milestone schedule, or bar chart.

Milestone Schedule

A schedule representing important events along the path to project completion. All milestones may not be equally significant. The most significant are termed "major milestones" and usually represent the completion of a group of activities.

Multiple Prime Contracts

Separate contractors contracting directly with the owner for specific and designated elements of the work.

Nonconforming Work

Work that does not meet the requirements of the contract documents.

Notice of Award

A formal document informing an individual or organization that it has successfully secured a contract.

Notice to Proceed

A formal document and/or point in the project's life cycle authorizing an individual or organization to commence work under its contract. The issuance of the notice to proceed typically marks the end of the procurement phase.

Owner Construction Management

A form of construction management that does not use an independent construction management organization as a team member. The owner performs all required construction management services with in-house staff.

Owner's Representative

The individual representing the owner on the project team.

Penalty

A punitive measure, usually associated with failure to fulfill a contractual obligation.

Performance Bond

A pledge from a surety guaranteeing the performance of the work or payment of the bond amount to the obligee (owner or contractor) in the event of a default in performance of contractual obligations.

Phased Construction

An incremental approach to construction or design and construction. Each overlapping or sequential phase or element has a defined work scope and is considered as a separate project.

Plans

See "Drawings."

Postconstruction Phase

The period following substantial completion.

Pre-Design Phase

The period before schematic design commences, during which the project is initiated and the program is developed; the planning and conceptual phase.

Prime Contract

A direct contract with an owner. It can be a single contract and/or include the work specified for several contracts, depending upon division of work.

Prime Contractor

A contractor who has a contract with an owner.

Professional Services

Services provided by a professional or by an organization that has specific competence in a field of endeavor that requires professional (and technical) knowledge and capabilities and that meets recognized standards of performance.

Program Management

The practice of professional construction management applied to a capital improvement program of one or more projects from inception to completion. Comprehensive construction management services are used to integrate the different facets of the construction process—planning, design, procurement, construction, and activation—for the purpose of providing standardized technical and management expertise on each project.

Progress Meeting

A meeting dedicated to the subject of progress during any phase of project delivery.

Progress Payment

Partial payment of the contract amount periodically paid by the owner, upon approval by the CM, verifying that portions of the work have been accomplished.

Project

The total effort required in all phases, from conception through design and construction completion, to accomplish the owner's objectives.

Project Budget

The sum or target figure established to cover all the owner's costs of the project. It includes the cost of construction and all other costs such as land, legal and consultant fees, interest, and other project-related costs.

Project Cost

The actual cost of the entire project.

Project Management

As applied to a construction project, the use of integrated systems and procedures by the project team to accomplish design and construction. Project management is an integral function of construction management.

Project Management Plan

A document prepared by the CM, and approved by the owner, that defines the owner's goals and expectations, including scope, budget schedule, and quality, and the strategies to be used to fulfill the requirements of the project.

Project Procedures Manual

A detailed definition of the project team responsibilities and authority, project systems, and procedures to be used for all phases of the project.

Project Team

Initially consists of the owner, designer, and CM. Thereafter, prime contractors are added to the team, as they are engaged for construction.

Project Team Meeting

A meeting dedicated to all aspects of the project, involving the project team members (owner, designer, CM, and contractor(s)).

Punch List

A list made near the completion of the construction work, indicating items of work that remain unfinished, do not meet quality or quantity requirements as specified, or are yet to be performed and must be accomplished by the contractor prior to completing the terms of the contract.

Quality

The degree to which the project and its components meet the owner's expectations, objectives, standards, and intended purpose; determined by measuring conformity of the project to the plans, specifications, and applicable standards.

Quality Assurance (QA)

The application of planned and systematic methods to verify that quality control procedures are being effectively implemented.

Quality Control (QC)

The continuous review, certification, inspection, and testing of project components, including persons, systems, materials, documents, techniques, and workmanship, to determine whether or not such components conform to the plans, specifications, applicable standards, and project requirements.

Quality Management

The process of planning, organization, implementation, monitoring, and documenting of a system of policies and procedures that coordinate and direct relevant project resources and activities in a manner that will achieve the desired quality.

Record Drawings

Drawings (plans), prepared after construction is complete, that represent the work accomplished under the contract.

Recovery Schedule

The schedule that depicts action(s) and special effort(s) required to recover lost time in the approved schedule. It can depict activities of any member of the project team.

Request for Change Proposal

A written document issued by the CM to the contractor that describes a proposed change to the contract documents, for purposes of establishing cost and time impacts.

May also be known as a bulletin or request for quote.

Schedule of Values
A list of basic contract segments, in both labor and material, where each line item consists of a description of a portion of work and a related cost, and the sum of the line items equals the total contract price. Generally used to determine progress payments to the contractor(s).

Scope
Identification of all requirements of a project or contract.

Scope Changes
Changes that expand or reduce the requirements of the project during design or construction.

Shop Drawings
Drawings typically prepared by the contractor, based upon the contract documents and provided in sufficient detail to indicate to the designer that the contractor intends to construct the referenced work in a manner that is consistent with the design intent and the contract documents.

Short-Term Construction Activity Plan
The planning and scheduling of prime contractor(s) activities on site, for the short duration or "foreseeable future," usually developed on a week-by-week basis, using milestones for planning intervals coordinated by the CM. Also known as a rolling schedule, "look ahead" schedule, or short-interval schedule.

Special Conditions (of the Contract for Construction)
See "Supplementary General Conditions."

Special Consultants
The designation for various professionals, including engineers, architects, designers, and other experts, who provide expertise in specialized fields.

Specifications
The detailed written descriptions of materials, equipment, systems, required workmanship, and other qualitative information pertaining to the work.

Start-Up
The period prior to occupancy when systems are activated and checked out, and the owner's operating and maintenance staff assumes the control and operation of the systems.

Subcontractor

A contractor who has a contract with a prime contractor to perform work.

Submittals

Transmittals of information as required by the contract documents.

Substantial Completion

The date, certified by the designer or CM or both, on which the contractor has reached the stage of completion when the facility may be used for its intended purposes, even though all work is not completed.

Supplementary General Conditions

Additions and/or modifications to the general conditions that are part of the bid documents and/or contract documents.

Testing

The application of specific procedures to determine if work has been completed in the prescribed manner and at the required levels of workmanship. See "Nonconforming Work."

Trade Contractors

Construction contractors who specialize in providing and/or installing specific elements of the overall construction requirements of a complete project.

Trade-Off Study

The study to define the comparative values and risks of a substitution or exchange of a design component. The trade-off can identify both monetary and functional values. Also known as an alternatives analysis.

Value Analysis

See "Value Engineering."

Value Engineering

A specialized cost control technique, which utilizes a systematic and creative analysis of the functions of a project or operation to determine how best to achieve the necessary function, performance, and reliability at the minimum life-cycle cost.

Warranty

Assurance by a party that it will assume stipulated responsibility for its own work.

Work

The construction, to include all labor, material, and equipment, required by the contract documents.

CMAA Student Chapters

ARIZONA

ITT Technical Institute, Tempe

5005 S. Wendler Drive
Tempe, AZ 85282
Host Regional Chapter: Arizona Chapter
Faculty Advisor: Instructor Alex Devereux
Phone: 602–437–7554
Email: adevereux@itt-tech.edu

CALIFORNIA

California Polytechnic State University, San Luis Obispo

1 Grand Avenue
San Luis Obispo, CA 93407
Website: www.construction.calpoly.edu/
Host Regional Chapter: Southern California Chapter
Faculty Advisor: Lonny Simonian
Phone: 805–756–7981
Email: lsimonia@calpoly.edu

California State University, Chico

Construction Management Department
400 West First Street
Chico, CA 95929-0001
Website: www.cmcsuchico.edu
Host Regional Chapter: Northern California Chapter
Faculty Advisors: Dr. James O'Bannon and Professor David Shirah
Phone: 530-898-5216
Email: O'Bannon (jim@rhainc.com) and Shirah (dshirah@csuchico.com)

California State University, Fresno

2320 East San Ramon Avenue, M/S EE94
Fresno, CA 93740
Website: www.csufresno.edu/engineering
Host Regional Chapter: Northern California Chapter
Faculty Advisor: Asst. Professor Brad Hyatt
Phone: 559-278-7735
Email: bhyatt@csufresno.edu

California State University, Long Beach

1250 Bellflower Boulevard
Long Beach, CA 90840
Website: www.csulb.edu/colleges/coe/cecem/
Host Regional Chapter: Southern California Chapter
Faculty Advisor: Dr. Tariq Shehab
Phone: 562–985–1643
Email: shehab@csulb.edu

California State University, Northridge

18111 Nordhoff Street, Suite JD4507
Northridge, CA 91330–8347
Host Regional Chapter: Southern California Chapter
Faculty Advisor: Associate Professor Mohamed Hegab, PhD
Phone: 818–677–7034
Email: mhegab@csun.edu

San Diego State University

Civil, Construction & Environmental Engineering
5500 Campanile Drive
San Diego, CA 92182
Website: www.engineering.sdsu.edu/civil/
Host Regional Chapter: San Diego Chapter
Faculty Advisor: Assistant Professor Thais Alves
Phone: 619–594–8289
Email: talves@mail.sdsu.edu

University of Southern California

3620 S. Vermont Avenue, KAP222
Los Angeles, CA 90089–2531
Host Regional Chapter: Southern California Chapter
Faculty Advisor: Professor and CE&M Program Director Henry Koffman
Phone: 213–740–0556
Email: koffman@usc.edu

Westwood College, South Bay

19700 Vermont Avenue, Suite 100
Torrance, CA 90502
Website: www.thecmclub.com
Host Regional Chapter: Southern California Chapter
Faculty Advisor: Adjunct Faculty Thomas Kempton
Phone: 310-855-8744
Email: tskempy@aol.com

COLORADO

Colorado State University

Department of Construction Management
Guggenheim Hall
Fort Collins, CO 80523–1584
Website: www.cahs.colostate.edu/cmaa/mission.html
Host Regional Chapter: Colorado Chapter
Faculty Advisor: Assistant Professor Mehmet Egemen Ozbek, PhD
Phone: 970–491–4101
Email: meozbek@cahs.colostate.edu

FLORIDA

Everglades University, Orlando

877 E. Altamonte Drive
Altamonte Springs, FL 32701
Website: www.evergladesuniversity.edu/newsite/degree-programs_prospective_students.asp#2
Host Regional Chapter: West Central Florida Chapter
Faculty Advisor: Lead Faculty Member Daniel Vannoy, JD, MSIR
Phone: 407–277–0311
Email: dvanoy@cfl.rr.com

University of Florida

M. E. Rinker, Sr. School of Building Construction
314 Rinker, P.O. Box 115703
Gainesville, FL 32611–5703
Website: www.bcn.ufl.edu/
Host Regional Chapter: West Central Florida Chapter
Faculty Advisor: Professor Michael Cook, MBA, JD
Phone: 352–273–1156
Email: zekecook@ufl.edu

GEORGIA

Southern Polytechnic State University

1100 S Marietta Parkway, Suite H-338
Marietta, GA 30060–7896
Website: www.spsu.edu/cnst
Host Regional Chapter: South Atlantic Chapter
Faculty Advisor: Assistant Professor Pavan Meadati, PhD
Phone: 678–915–3715
Email: pmeadati@spsu.edu

ILLINOIS

Illinois Institute of Technology

Construction Engineering and Management Program
Department of Civil Architectural and Environmental Engineering
3201 South Dearborn Street
Alumni Memorial Hall, Room 229
Chicago, IL 60616
Website: www.iit.edu/%7Ecmaa
Host Regional Chapter: Chicago Chapter
Faculty Advisor: Professor and CE&M Director David Arditi, PhD
Phone: 312–567–3546
Email: Arditi@iit.edu

Southern Illinois University-Carbondale

College of Applied Sciences & Arts
1365 Douglas Drive, Mailcode 6614
Carbondale, IL 62901–6614
Website: www.siu.edu/
Host Regional Chapter: Chicago Chapter
Faculty Advisor: Assistant Professor J. Kevin Roth
Phone: 618–453–7219
Email: jkroth@siu.edu

University of Illinois, Urbana-Champaign

205 North Mathews Avenue
3129b Newmark Engineering Lab
Urbana, IL 61801–2352
Host Regional Chapter: Chicago Chapter
Faculty Advisor: Associate Professor Liang Y. Liu
Phone: 217–333–6951
Email: LLiu1@illinois.edu

INDIANA

Purdue University

Civil Engineering & Construction Engineering and Management
550 Stadium Mall Drive, Civil 1227
West Lafayette, IN 47907-2051
Website: www.engineering.purdue.edu/cem
Host Regional Chapter: Indiana Chapter
Faculty Advisor: Asst. Professor Amr Kandil
Phone: 765-494-2246
Email: akandil@purdue.edu

MAINE

University of Southern Maine

Department of Technology
37 College Avenue
Gorham, ME 04038
Website: www.usm.maine.edu/tech/programs/const_magt.html
Host Regional Chapter: New England Chapter
Faculty Advisor: David Early
Phone: 207–780–5440
Email: dearly@usm.maine.edu

MASSACHUSETTS

Wentworth Institute of Technology

550 Huntington Avenue
Boston, MA 02115
Host Regional Chapter: New England Chapter
Faculty Advisor: Associate Professor E. Scott Sumner, CCM
Phone: 617–989–4259
Email: sumnere@wit.edu

MINNESOTA

Minnesota State University Moorhead, Twin Cities

7411 Eighty-Fifth Avenue North
Brooklyn Park, MN 55445
Website: www.mnstate.edu
Host Regional Chapter: Minnesota Chapter
Faculty Advisor: Assistant Professor Robert Riesselman, PE
Phone: 763-488-0495
Email: robert.riesselman@mnstate.edu

NEW YORK

Columbia University School of Continuing Education

Construction Management Department
510 Lewisohn Hall, 2970 Broadway MC 4110
New York, NY 10027
Website: www.ce.columbia.edu/construction
Host Regional Chapter: Metropolitan New York/New Jersey
Faculty Advisor: Industry Liaison Jamie Daniels
Phone: 914-574-1373
Email: jd2564@columbia.edu

New York Institute of Technology, Manhattan

1855 Broadway, 11th Floor
New York, NY 10023
Website: www.nyit.edu/
Host Regional Chapter: Metropolitan New York/New

Jersey Chapter
Faculty Advisor: Associate Dean Frank Mruk
Phone: 212-261-1676
Email: fmruk@nyit.edu

New York Institute of Technology, Old Westbury

Northern Boulevard, P.O. Box 8000
Old Westbury, NY 11568
Website: www.nyit.edu/
Host Regional Chapter: Metropolitan New York/New Jersey Chapter
Faculty Advisor: Associate Dean Frank Mruk
Phone: 212-261-1676
Email: fmruk@nyit.edu

New York University, Schack Institute of Real Estate

11 West 42nd Street, Room 509
New York, NY 10036
Website: www.scps.nyu.edu/areas-of-study/real-estate/graduate-programs/ms-construction-management/
Host Regional Chapter: Metropolitan New York/New Jersey Chapter
Faculty Advisor: Clinical Associate Professor Richard Lambeck, PE
Phone: 212-992-3219
Email: RL79@nyu.edu

Polytechnic Institute of New York University

6 Metro Tech Center
Brooklyn, NY 11201
Host Regional Chapter: Metropolitan New York/New Jersey Chapter
Faculty Advisor: Professor Lawrence Chiarelli
Phone: 718-260-4040
Email: lchiarel@poly.edu

Pratt Institute

144 West 14th Street
New York, NY 10011
Host Regional Chapter: Metropolitan New York/New Jersey Chapter
Faculty Advisor: Department Chair Harriet Markis
Phone: 212-647-7524
Email: pratt.cmstudentchapter@gmail.com

NORTH CAROLINA

North Carolina A&T State University

Construction Management & Occupational Safety and Health
1601 East Market Street, 110 Price Hall
Greensboro, NC 27411
Website: www.ncat.edu
Host Regional Chapter: North Carolina
Faculty Advisor: Department Chair Dr. Robert Pyle
Phone: 336-334-7590
Email: pyler@ncat.edu

OHIO

Kent State University

College of Technology
209A Van Deusen Hall
Kent, OH 44242
Website: www.ksucm.com
Host Regional Chapter: Ohio Chapter
Faculty Advisor: Assistant Professor Joe Karpinski
Phone: 330-672-3080
Email: jkarpins@kent.edu

PENNSYLVANIA

Drexel University

3001 Market Street, Suite 173
Philadelphia, PA 19104
Host Regional Chapter: Mid Atlantic Chapter
Faculty Advisor: Assistant Clinical Professor Stanley Jackson, CCM, CDT, LEED AP, PhD Candidate
Phone: 215-895-5966
Email: slj28@drexel.edu

Temple University

1801 North Broad Street
Philadelphia, PA 19122
Host Regional Chapter: Mid Atlantic Chapter

Faculty Advisor: Dr. Philip Udo-Inyang
Phone: 215–204–7831
Email: philip.udo-inyang@temple.edu

TEXAS

Texas A&M University, College Station

Department of Construction Science, College of Architecture
3137 TAMU, Langford Building A, Room 422
College Station, TX 77843
Website: http://cmaa.tamu.edu/
Information: info@cmaa.tamu.edu
Host Regional Chapter: South Central Texas Chapter
Faculty Advisor: Associate Professor Julian Kang, PhD
Phone: 979–845–1017
Email: juliankang@tamu.edu

Westwood College, Houston South

7322 Southwest Freeway
Arena Tower One
Houston, TX 77074
Website: www.westwood.edu
Faculty Advisor: Ruhina Surendran
Phone: 832-316-0424
Email: rsurendram@westwood.edu

VIRGINIA

Virginia Tech

200 Patton Hall
Blacksburg, VA 24060
Website: www.cmaa.org.vt.edu/
Host Regional Chapter: National Capital Chapter
Faculty Advisor: Vecellio Professor of CE&M Jesus de la Garza, PhD
Phone: 540–231–5789
Email: chema@vt.edu

Books and Publications available from CMAA at www.cmaanet.org

Bibliography

***Construction Management Standards of Practice,* 2010 Edition**

The comprehensive new edition of CMAA's SOP includes new sections on Risk Management, BIM, and Sustainability, along with expanded attention to Program Management and other updates. This is the basic work defining the range of services that constitute professional construction management, serving as an indispensable guide for owners and service providers alike.

Program Management: Concepts and Strategies for Managing Capital Building Programs

Chuck Thomsen, FAIA, FCMAA
2008, 303 pages

The industry's first in-depth look at a new approach to managing ongoing building programs—a new strategy that can produce lower costs, shorter schedules, lightened administrative burdens, and significant improvements in the quality of finished product. Chuck Thomsen, the first person to become a fellow of both AIA and CMAA, shares the insights of a half-century of experience and innovation.

Program Management 2.0: Concepts and Strategies for Managing Capital Building Programs (Revised)

Chuck Thomsen, FAIA, FCMAA, and Sid Sanders
2011, 423 pages

The co-authors take a fresh look at the concepts presented In *Program Management* from the viewpoints of both a consultant and an owner/client, incorporating new IT developments, market trends, and feedback from presentations and discussions nationwide.

Strategic Program Management

Bob Prieto
2009, 124 pages

Strategic Program Management stresses the close connection between successful program management and the owner's business strategy. "In its simplest form an organization's strategic business objectives are addressed through development of a comprehensive strategic plan. Program Management is about translating that strategic plan into a defined set of discrete but inter-related activities and then managing the delivery and successful completion of these activities in a holistic way," the author notes in his introduction.

Program Management in the Engineering & Construction Industry

Bob Prieto
2011, 408 pages

Prieto's book reflects continued research and work on those attributes of large engineering and construction programs that drive success, as well as those that present challenges to owners and their program managers. This work builds on his first book and incorporates later thinking as reflected in numerous papers and presentations.

Building Tall: My Life and the Creation of Construction Management

John L. Tishman and Tom Schachtman
2010, 226 pages

John Tishman recounts his experiences in the formative years of the CM profession, including his role as CM on some of the world's most famous and complex projects: New York's World Trade Center, Chicago's Hancock Tower, Madison Square Garden, EPCOT Center at Walt Disney World, and many others. He expresses his views concerning where CM fits in today's complex construction environment and what values it can bring to contemporary programs.

Available from John Wiley & Sons

Construction Project Management: A Practical Guide to Field Construction Management, 5th Edition

S. Keoki Sears, Glenn A. Sears, and Richard H. Clough
2008, 408 pages

Long considered the preeminent guide to the Critical Path Method (CPM) of project scheduling, *Construction Project Management* by Clough and Sears combines a solid foundation in the principles and fundamentals of CPM with particular emphasis on project planning, demonstrated through an example project.

This fifth edition features a range of improvements. New pedagogical devices improve absorption of the material. Updated labor, material, and equipment pricing is incorporated into the text. Coverage is enhanced by discussions of contemporary planning and management methods such as Work Breakdown Structures (WBS) and the Earned Value Management System (EVMS).

The fifth edition features include:

- Complete coverage of planning and scheduling principles that apply to every type of construction project
- Expanded coverage of production planning
- Large foldout illustrations conveniently integrated throughout the book

Useful Web Resources

American Association of State Highway and Transportation Officials

www.aashto.org

American Council for Construction Education

www.acce-hq.org

American Council of Engineering Companies

www.acec.org

American Institute of Architects

www.aia.org

American Institute of Constructors

www.aicnet.org

American Public Transportation Association

www.apta.org

American Public Works Association

www.apwa.net/

American Road & Transportation Builders Association

www.artba.org

American Society of Civil Engineers

www.asce.org

Associated Builders and Contractors

www.abc.org

Associated General Contractors

www.agc.org

Certified Construction Manager (CCM®) Credential

www.cmcertification.org

Complete information on the qualifications and procedures for obtaining the industry's most respected professional credential.

Chartered Institute of Building

www.ciob.org.uk/home

CM Career and Educational Information

www.cmaanet.org/cmaa-foundation

The CMAA Foundation supports research, career promotion and information, and scholarships for CM students.

CMAA Regional Chapters

http://cmaanet.org/cmaa-chapters

Chapters listed by territory, with links to chapter websites.

Construction Financial Management Association

www.cfma.org

Construction Industry Institute

www.construction-institute.org

Construction Industry Round Table

www.cirt.org

Construction Management Association of America

www.cmaanet.org

The Construction Management Association of America is North America's only organization dedicated exclusively to the interests of professional construction and program management.

Construction Owners Association of America

www.coaa.org

Construction Specifications Institute

www.csinet.org

International Construction Project Management Association

www.icpma.net

International Facility Management Association

www.ifma.org

National Association of Women in Construction

www.nawic.org

National Institute of Building Sciences

www.nibs.org

National Society of Professional Engineers

www.nspe.org.

Professional Women in Construction

www.pwcusa.org

Scholarship Information from CMAA Foundation

http://cmaanet.org/cmaa-foundation-scholarships

Society of American Military Engineers

www.same.org

Society of Value Engineers

www.value-eng.org

Sustainable Buildings Industry Council

www.sbicouncil.org

U.S. Green Building Council

www.usgbc.org

Women's Transportation Seminar

www.wtsinternational.org